CROSS CURRENTS

Also by the author:

MECHANISMS OF GROWTH CONTROL
ELECTROMAGNETISM AND LIFE
THE BODY ELECTRIC (with Gary Selden)
THE HEALTH HAZARDS OF ELECTROMAGNETIC RADIATION

CROSS CURRENTS

The Promise of Electromedicine
The Perils of Electropollution

ROBERT O. BECKER, M.D.

Jeremy P. Tarcher / Penguin
a member of
Penguin Group (USA) Inc.
New York

Most Tarcher/Penguin books are available at special quantity
discounts for bulk purchases for sales promotions, premiums,
fund-raising, and educational needs. Special books or book excerpts
also can be created to fit specific needs.
For details, write Penguin Group (USA) Inc.
Special Markets, 375 Hudson Street,
New York, NY 10014

TO MY WIFE LILLIAN

Jeremy P. Tarcher / Penguin
a member of
Penguin Group (USA) Inc.
375 Hudson Street
New York, NY 10014
www.penguin.com

Library of Congress Cataloging-in-Publication Data

Becker, Robert O.
    Cross currents : the promise of electromedicine, the perils of
electropollution / Robert O. Becker.
        p.   cm.
    Includes index.
    1. Electromagnetic waves—Health aspects.   2. Electrophysiology.
    3. Biomagnetism.   I. Title.
QP82.2.E43B43   1990
612'01427—dc20   89-5067
ISBN 0-87477-609-0   CIP

Design by Susan Shankin
Computer art by Marjory Dressler

Manufactured in the United States of America

CONTENTS

PART TWO: ELECTROMAGNETIC MEDICINE

PART THREE: ELECTROMAGNETIC POLLUTION

ACKNOWLEDGMENTS

I wish to thank first of all my editor, Connie Zweig. This book would not have been written without her. Her patience and tact are responsible for a readable, understandable story. Thanks are due also to all those who helped, encouraged, and provided information and guidance; they are too numerous to mention individually. A special thanks to my son, Adam, for lengthy and numerous discussions and for his skill at computer graphics, which produced simple, clear illustrations from my complex scrawls. And, finally, thanks to The Institute of Noetic Sciences, for invaluable assistance and support.

PREFACE

*D*r. Robert Becker's prescient book, *Cross Currents*, was published twelve years ago and has even more resonance today as our environment becomes increasingly saturated with electromagnetic signals. Today, some 150 million Americans—and more than a billion others worldwide—regularly use mobile phones, irradiating their brains and their eyes.

Because of the great demand for cell-phone service, we are surrounded by towers and base stations. The ambient microwave levels in urban areas are now ten times higher than in the pre-wireless age, and we can expect further increases as we begin to use handheld devices to send pictures and information.

Will we see an increase in brain cancer or neurological disease decades from now? Despite reassurances from industry and government officials, the answer is that we do not know. Nor are we likely to find out, unless the current research effort is stepped up.

The experience with power-line electromagnetic fields (EMFs) strongly suggests that we should be paying more attention to the possible health impacts of mobile phones. When Dr. Becker was writing this book, evidence was accumulating that power-line EMFs are linked to cancer. Though the electric utility industry denounced this association for years, there is now an international consensus among epidemiologists that low levels of EMFs can double the risk of childhood leukemia. (The companies continue to be in denial.)

In the summer of 2001, an expert panel convened by the International Agency for Research on Cancer (IARC) unanimously classified EMFs as a "possible human carcinogen," with some panel members arguing that the scientific evidence warrants an even stronger public warning.

Of course, this finding begs the question: *If such fields can promote cancer among children, what else can they do?* Here again, we are in uncertain territory with no reliable answers.

Progress comes slowly. It took more than twenty years for Dr. Nancy Wertheimer and Ed Leeper's EMF–childhood cancer hypothesis—first published in 1979— to win official recognition by IARC. We are now confronted with similar early warnings about mobile phones, radar, and radio and television towers, but we may have to wait another generation before we finally understand the nature and extent of these possible microwave health risks.

Fortunately, a handful of dedicated researchers are following in Dr. Becker's footsteps and are trying to work out how weak fields can influence living systems. However, as Dr. Becker makes clear here in *Cross Currents* and in his earlier classic, *The Body Electric*, they face an uphill battle.

Dr. Becker has faith that an informed public is the key to the future, and I agree with him. Public pressure prompted the power-line studies that led to the IARC warning. If the public continues to demand the truth about mobile phones and other sources of electromagnetic radiation, the studies will be performed.

One union official in Sweden, almost single-handedly, forced computer manufacturers to change the way video display terminals are designed. He showed that their electromagnetic emissions could be reduced at a negligible cost. As a result, millions of computer users in offices and schools or at home need no longer be concerned about VDT radiation.

As Dr. Becker points out, the perils of electropollution and the promise of electromedicine are two sides of the same issue. The sooner we learn about the biological effects of electromagnetic signals, the sooner we can harness them to fight disease.

We are at the cusp of an amazing electromagnetic revolution. It promises to be exciting and rewarding—if we make the necessary effort to explain what these fields can and cannot actually do.

LOUIS SLESIN
Editor, *Microwave News*
New York City
July 2003

PROLOGUE

*C*ross Currents describes the meeting of two opposing trends: the rapid rise of electromedicine, which promises to unlock the secrets of healing, and the parallel rise of electropollution, which poses a pressing environmental danger.

The book explains that current popular healing practices use an invisible common source: the body's innate electrical systems. While this discovery is coming to light, *Cross Currents* reports evidence that these same bodily resources are being adversely affected by man-made electromagnetic fields from widespread technologies.

The promise of electromagnetic healing from methods such as acupuncture, hypnosis, homeopathy, visualization, psychic healing, and electrotherapy is dampened by the perils of electromagnetic pollution from devices such as power lines, radar, microwaves, satellites, ham radios, and even electric appliances.

The story has its roots in the history of medicine. The twin successes of science and technology of World War II, the atom bomb and penicillin, promised us a new world. In that world we would attain complete control over our environment, have access to free energy for our homes and cars, and enjoy freedom from disease—all through further advances in scientific knowledge. Scientific research became a national objective. Institutions such as the National Institutes of Health and the National Laboratories were established, and the funds poured in.

During the past forty years, our world has been shaped by these concepts of "big science" and "big technology." Initially, the promise seemed to be fulfilled: successes were immediate, and we looked forward to continuing progress.

However, things have changed dramatically. We now realize that our global ecosystem is collapsing, that energy is far from free, and that medically we are little—if at all—better off than we were in 1950. We have defeated the great epidemics of the past, only to see equally pernicious plagues emerge.

How could this have happened? Where did we go wrong, and what can we do? The scientific enterprise that started in the 1950s with such excitement and promise has grown into a scientific establishment that appears unable to cope with today's problems, except to advocate more and more of the same technology. The problem is not with science itself but with the fact that science is a human endeavor. Scientists are not always logical seekers of the truth, as they are portrayed in the popular press, but human beings subject to the same failings as the rest of humanity.

The modern career scientist's business is one in which success is measured by the number of papers published. Maximizing this number leads to greater prestige, more grant funds, larger laboratories, and positions on decision-making committees. Unfortunately, it is much easier to get a paper published if it does not challenge the present orthodoxy. As a result, few career scientists are willing to look at issues that cast doubt on established beliefs. Rather, they are content to mull over the few questions that remain from discoveries of the past. These factors lead to stagnation and to the appearance that scientific progress is the slow, painstaking acquisition of small bits of knowledge that are simply adding on, in a minor way, to an edifice that is already extant. For the most part, science today has lost its most essential aspect—the spirit of adventure.

As Thomas Kuhn showed so well, the path of science is characterized by periods of revolution during which the dogmatic system of beliefs derived from the past is replaced by a new paradigm—a new view of how things work. This change is brought about by the gradually increasing failure of the established paradigm to deliver on its promises and the ability of the new one to better explain reality. As in all revolutions, the change in ideas is always bitterly opposed by the practitioners of the established paradigm.

The paradigm that was in place in 1950 was based on the chemical-mechanistic concept of life. In this view, living things were chemical-mechanical machines whose capabilities were constrained to those functions permitted by this model; there was no place for any characteristics, such as autonomy or self-healing, that did not fit this mold. This view was reinforced until it became a dogma, the proponents of which claimed to know everything there was to know about life. This paradigm not only dominated our society but ruled the medical profession as well, limiting both the methods that could be used to bring about a cure and our perception of the ability of the human body to heal itself.

As each technological advance entered medical practice, we found ourselves paying an increasing price of unexpected side effects. Most technological cures for cancer, for example, were found to be carcinogenic themselves. Because such unexpected side effects have required additional "technological fixes," we now find ourselves in a spiral in which technological applications are piled one atop another, with no end in sight and no cure for the patient.

The chemical-mechanistic paradigm has failed, and a medical revolution has begun. Today, motivated by a growing dissatisfaction with the mechanistic concept and the treatments it dictates, many physicians are reexamining and applying therapeutic techniques that were previously discarded as "unscientific" by academic medicine. The use of foods, herbs, meditation, and acupuncture are only a few examples. This radical change in medical practice is deeply rooted in ancient concepts of life, energy, and medicine, and it includes a reaffirmation of the innate healing ability of living things.

Simultaneously, the integration of physics and biology has given rise to a new scientific revolution, which has revealed unexpected complexities and previously undreamed-of abilities in living things. The chemistry of life has been revealed to be based upon the underlying forces of electricity and magnetism. Our bodies and brains generate electromagnetic fields within us and around us. I began to tell this story in my book *The Body Electric* (Morrow, 1985). Since then, the pace of research done around the world has increased, and even more significant findings are being reported.

Not only does this new vision broaden the scope of our biological capabilities, but it also relates living things to the factors of electricity and magnetism in our global environment. After all, we live in the Earth's natural magnetic field, and we have created a vast global network of man-made electromagnetic fields. Indeed, life today can be seen as fields within fields within fields.

These revolutionary changes in science and medicine have been on parallel but unconnected paths. When the two are brought together, their synthesis results in a totally new paradigm. Ancient concepts of life and its relationship to the forces of the universe are shown to be correct in many ways. The mind can influence the body, and the body has an innate self-healing system. The results of alternative medical practices—acupuncture, homeopathy, hypnosis, healing, electromedicine—can at last be seen to have a known mechanism of action.

This emerging paradigm will markedly change the future prac-

tice of medicine, and it will provide more effective and safer therapies. But it will also reveal ominous changes in disease patterns,
including the appearance of new diseases that have resulted from
our unlimited use of electromagnetic energy for power and communications. The reintroduction of electricity and magnetism into science and medicine will change forever the way we view ourselves
and our relationship to the global environment.

The phenomenon of life is controlled by the same forces that
have shaped the universe. From the beginning, life has been dependent on the Earth's natural electromagnetic environment. Today
this natural environment is submerged beneath a torrent of electromagnetic fields that have never before been present. In my previous
book I told the story of the human body electric. In *Cross Currents*
I will show how both the human body electric and the Earth's body
electric have been damaged by this alteration; I will then explain
what steps we must take to prevent the disaster that is fast approaching.

In Part One, "Life Energy: The Hidden Dynamic in Medicine,"
you will read about the history of life energy in medicine, the origins
of this kind of medicine in ancient cultures, and the way in which it
was lost with the gradual development of scientific and technological
medicine. I then tell the story of the discoveries of electricity and
magnetism in the human body, which led to my proposal of a "dual
nervous system," an innate and unseen healing system within us.

Part Two, "Electromagnetic Medicine," tells how these discoveries provide a scientific basis for ancient and current alternative
techniques. You will read about hypnosis, healing, visualization,
placebos, homeopathy, acupuncture, and electrotherapy.

In Part Three, "Electromagnetic Pollution," I tie the Earth's
natural magnetic field around us to our internal electrical and magnetic systems. I describe the growth of the man-made electromagnetic environment, which is composed of radio and TV signals, microwave transmissions, high-power tension lines, radar, and other
electromagnetic radiation that crisscrosses the world. Many studies
have shown that this radiation, which was previously thought to be
innocuous, may be extremely hazardous to our health. In fact, it has
been linked to the increased incidence of certain cancers, birth defects, learning disabilities, and mood changes. I believe that we can
link this invisible, highly dangerous radiation to many of the new
diseases that have recently emerged. In chapter 11, I discuss AIDS,
chronic-fatigue syndrome, Alzheimer's disease, autism, cancer, and
sudden infant death syndrome in light of these new findings.

The final chapter tells what you can do in your home and office to reduce the danger of electromagnetic pollution. I discuss the risks and benefits of personal computers, fluorescent lights, microwave ovens, personal radio transmitters, and electric clocks, blankets, hair dryers, and heaters.

I am convinced that these innovative scientific ideas should not be restricted to the scientific priesthood, but rather should be made available in understandable form to the general public. Many important policy decisions will soon need to be made. They should be made by an informed public, not by politicians, bureaucrats, or scientists who are blindly obedient to the tenets of their faith.

On the one hand, this knowledge offers great promise about our capacities to heal ourselves; on the other hand, it signals a dire warning, much like Rachel Carson's *Silent Spring,* which revealed our ecological crisis and launched a global environmental movement. I believe that we are facing another invisible crisis, one that will call for concerted action by an informed public.

PART ONE

LIFE ENERGY:
THE HIDDEN DYNAMIC IN MEDICINE

CHAPTER ONE

THE HISTORY OF
LIFE, ENERGY, AND MEDICINE

Since the days of revelation, in fact, the same four corrupting errors have been made over and over again: submission to faulty and unworthy authority; submission to what it was customary to believe; submission to the prejudices of the mob; and worst of all, concealment of ignorance by a false show of unheld knowledge, for no better reason than pride.

ROGER BACON

*M*artha's story: We sat in a quiet corner of the lab, the best I could do on short notice. She had called a few minutes before, asking to see me as soon as possible. Because of the ring of urgency in her voice, I suggested that she come right over.

She said, "I want you to know that I'm twenty-two years old and capable of making up my own mind. Dr. X at the Medical Center tells me that I have leukemia, a kind that's hard to treat, and that the only chance I have is to take one of the new experimental drugs. It might work, but all my hair will fall out and it will make me very sick. Dr. Becker, I want to know what you would do if you were me."

I had to think this over, so instead of answering right away, I asked her enough questions to make sure that the diagnosis had been well established. She described the complete workup that she had gone through and produced a copy of her last bone-marrow report. She said that she hated the bone-marrow biopsies in particular because they were so painful.

She went on to say that she had watched her mother die in horrible discomfort from chemotherapy for abdominal cancer, and that she felt that the treatment had not only been useless but had destroyed her mother's last days. "I will *not* take the same kind of treatment," she said emphatically. "I've come to you because I've heard that you're different and can help me find some other treatment."

I assured her that I certainly agreed with her opinion about her mother's case. "But I don't see what I can do in your case," I told her. "You seem to have already made up your mind not to accept the recommended treatment. What do you expect me to be able to do?"

She had spirit, and she got mad. "I came to you because I thought you could advise me on what to do myself or could send me to somebody who could give me some other treatment. But you're just like all the rest. I guess I wasted my time!" She got up to leave, but I motioned her to sit back down.

I told her that a few months before, I had been at a meeting in another city and had met a well-qualified physician who had the courage to treat cancer patients through diet, visualization, and biofeedback techniques, with some very interesting results. "I agree with your decision," I said, "and I can send you to this doctor. However, I can make no promises, and I can't assure you that it will work. But, to be honest, that's what *I* would do."

She agreed to try it, and I made the phone call. She left, promising to keep in touch. I received a postcard about a month later: "Having a wonderful time—wish you were here!" Then I heard nothing more from her for almost a year, when she walked into my lab and announced, "I'm cured!"

She described the treatment she had undergone, and how she had progressed until her last three bone-marrow determinations were absolutely normal. She had waited until both she and the doctor were sure of the result before coming to see me. She then said, "I have to tell you something else. I just came from Dr. X's office. I told him what I had done and that I was cured, and I even offered to let him do another bone marrow to see for himself. He said, 'Get the hell out of here,' and he refused to even talk with me."

She sent me one more postcard about two years later. It read simply, "Still okay—Thanks."

Jerry's story:
Most of my unofficial patients were employees at the hospital or members of their families who would come to me with apologies

for "taking up my time." One day, one of the electricians was working on the wiring in the lab while I was eating my lunch. "Say, Doc, my son has been having some trouble lately," he suddenly said. "He seems to lose his balance sometimes. What could cause that?" A few questions established that the boy, who was only seven years old, also complained of dizziness, headaches, and of some problems with seeing, and that all of this was rapidly getting worse. The family doctor thought it might be an ear infection and had treated the child for this, but there had been no improvement.

The possibility of a brain tumor occurred to me immediately. I said, "Listen, this could be something pretty serious. It might even be something in his brain. I think you should take him to a good neurologist or neurosurgeon as soon as possible."

A few days later the electrician came into the lab and said, "Doc, you were right. We took Jerry to Dr. Y at the Medical Center, who ordered some tests and told us it might be a brain tumor." I called the doctor's office and found that a CAT scan had shown a mass in the back of the brain, and that exploratory surgery was scheduled for the next day. I told the father that I thought everything possible was being done, and that most of these tumors turned out to be benign.

The next afternoon, the father came in with a broad grin, saying, "You were right again. Dr. Y said the tumor looked benign. He took out almost all of it, and he thinks Jerry will be okay." I said that was good news, indeed. However, I still felt somewhat uneasy about this case. I told the father, "Be sure and let me know what the pathologist says it is." Four days later he came back, saying that the pathologist also reported a benign tumor. I breathed a sigh of relief and told the father that I was as pleased as he was. Jerry remained in the hospital, recovering, and all seemed well.

One day a week or so later, I arrived at work to find the electrician waiting for me at the lab door. "Last night, my wife was in the hospital bringing Jerry his clothes because he was supposed to come home this morning," he began, obviously upset. "One of the interns walked in and told her that Jerry had cancer of the brain and that they had to start X-ray treatment this morning." His eyes blurred with tears. "Doc, why in the hell didn't they tell us that before? My wife got all upset, and someone else had to call me at home to tell me to come in. We don't know what to do. I've seen brain-cancer patients after X-ray treatment, and I don't want Jerry to go through that. When we said we wanted to think it over, they sent us to social services, where they told us that if we don't sign the permission, they'll take Jerry away from us!" I assured him that he did not have to sign the permit, and that I wanted to see both him and his wife to talk it over.

I called a friend in the pathology department at the Medical Center. He gave me the diagnosis and said that the staff had sent the slides out for review to the Armed Forces Institute of Pathology, where experts had unanimously diagnosed the tumor as malignant. My friend could not explain how the diagnosis had been missed at the Medical Center. I then called a friend and neurosurgeon at the Mayo Clinic and explained the situation to him. He said that there was nothing that could be done for this kind of tumor. The X-ray treatment would slow its growth for a while, but at the price of nausea, pain, and discomfort. There was no chemotherapy that would do any good, so none should be given. The boy was going to die; if Jerry were his son, he would simply make him comfortable and help him to enjoy his few remaining months. I asked him for his opinion on such alternative treatments as diet and visualization. He said he thought these were nonsense, but he agreed that such treatments would do no harm. His main objection was that they might give the parents "false hope," making it harder for them to accept the inevitable.

When I talked to the electrician and his wife I told them what this neurosurgeon had said. I told them that they had the authority to refuse any treatment, and that if this was their choice I would help them. I then described the alternative treatments, saying, "They're not approved, but they do no harm, and sometimes they work. If Jerry were my son, that's what I would do. If you'd like, I can contact some doctors you could go to."

The father was willing, but his wife totally rejected the idea, so I did not pursue it further. They signed the permission papers for X-ray treatment, and the father kept me posted on little Jerry. The boy finished the X-ray treatment and finally got over the nausea, but had lost a lot of weight and was quite weak. A few weeks later, he began having headaches again, and a CAT scan showed that the tumor was growing. The doctors recommended chemotherapy with an experimental drug that they said "might work." Jerry would have to go to a hospital more than a hundred miles away, but this was, the doctors said, his "last chance."

Jerry and his mother made the trip, and the father visited on the weekends. Every Monday, the father would report his son's progress to me. The chemotherapy was very bad, and Jerry was very sick. His hair had fallen out; he had lost more weight; he couldn't eat and had to be fed intravenously. He screamed every time the chemo-therapy was given.

The electrician said, "Doc, it tears my heart out. All he asks for is to have them stop sticking him with needles and giving him the medicine that makes him so sick so he can go home." All I could do for the father was sympathize with him.

Then one Sunday, I received a phone call from him: "We just brought Jerry home. Can we come over and see you?" They arrived, with Jerry propped up on pillows in the back seat of the car. He was a very sick little boy, not much more than skin and bones; he was breathing with difficulty and was unable to move because of the pain. His parents and I moved away from the car a little, and his mother said, "I would watch the nurses give him the chemo every day, and I just assumed they were doing it right. Yesterday, they left the empty bottle on the bedside table and I looked at it. I found that all along they had been giving him *twice* the dose he was supposed to get. I asked to see the doctor right away, and when he came in I told him what I had found. At first he said the dose was double because Jerry had such a bad cancer. Finally, he admitted it had been a mistake. But he said, 'It doesn't make any difference—the boy is going to die anyway.' " The mother immediately signed Jerry out of the hospital, against medical advice, and brought him home.

She said, "I don't know—maybe you were right, and we should have taken your suggestion. I guess there's nothing you can do for him now." I agreed that nothing could be done now, explaining that Jerry's immune system and other defenses had been destroyed by the X-ray treatments and chemotherapy. I made sure that the doctors had given them some effective narcotics for the pain, and we went back to the car. Very gently I shook Jerry's hand, not saying anything. He looked up at me. He knew. Two weeks later, Jerry died.

These two case reports are actually combinations of my experiences with many different individuals. However, I have encountered all of the singular events described at one time or another in my thirty years of practice. The two stories above illustrate the division of present-day medicine into two seemingly incompatible philosophies, "orthodox" medicine and "energy" medicine. The proponents of orthodox medicine—the kind taught in medical school and promoted by the American Medical Association (AMA) and the Food and Drug Administration (FDA)—are absolutely convinced that the body is simply a machine that cannot heal itself, and that the only appropriate therapies are powerful drugs and mechanical technologies. Proponents of "energy" medicine, on the other hand, believe that the body is more than a machine, and that it is capable not only of healing itself but also of performing other actions that lie completely outside the realm of established science. The latter practitioners believe that an appropriate therapy is one that either encourages the body's own energetic systems or that adds external energy to these systems.

Energy medicine is actually a generic term, one that is used to refer to many ancient and some modern medical practices. It is derived from the concept of alternative or holistic medicine, which encompasses many different therapeutic disciplines but which views each as a separate entity with different mechanisms of action.

The idea of an energy medicine that postulates a common basis for all of these disciplines has arisen lately because of new discoveries in biology that cast serious doubt on strictly mechanistic concepts of life and seem to reintroduce the old idea of a body energy. These discoveries are not "vitalistic" in the sense that they prove or even hint at some mysterious, unknowable life force. Rather, they reveal the presence of electrical and magnetic forces at the most basic levels within living organisms. These discoveries, which are the basis for a new scientific revolution, are rapidly changing what we believe about the way living things work.

The practitioners of each discipline of alternative medicine (for example, acupuncturists, homeopaths, and electrotherapy practitioners) eagerly—and often prematurely—embraced this latest scientific revolution as their own, postulating that these energetic systems were the ones accessed by their particular techniques and that other alternative methods lacked this ability. However, more recently the idea has arisen that this one common principle could be applied to all alternative medical techniques. As a result, the generic term *energy medicine* was coined.

While this scientific revolution was brewing, a revolution in medical practice was also occurring. Patients were becoming increasingly dissatisfied with orthodox medicine. The promises of technology implied by such ideas as "the war on cancer" were unfulfilled. That war seemed, in fact, to have turned into one of attrition, with the patients suffering the casualties and bearing the increasing expenses. Hospitals were becoming dangerous places to enter; patients sometimes entered with minor illnesses and left with permanent disabilities resulting from one complication after another. The old adage, "The doctor knows best" began to be replaced with "The doctor is usually wrong" or, worse, "Your doctor can make you sicker than you already are."

Some patients discovered the various disciplines of energy medicine, which appeared to have three outstanding things to offer. First, they would do no harm; second, they often seemed to do some good; and third, they were much less expensive than orthodox medicine. As a result, the number of patients treated with these unorthodox, unapproved techniques has increased markedly over the past decade.

In sum, the revolution in biology and medicine has been brought about in part by consumer dissatisfaction with orthodox medicine—with its increasingly impersonal technological sophistication, its high costs, and its inefficiency—and in part by the new scientific data challenging the mechanistic view of life.

There are many questions raised by this revolution. Is energy medicine scientifically real, or is it simply the newest quackery? Which of the numerous techniques, if any, holds the key to its application? Are there really bodily energies that play a role in healing? And, most importantly, will energy medicine survive to dominate orthodox medicine or to become integrated with it? Or will it simply disappear, like so many fads of the past?

Before we even attempt to bring some order to the present disorder, we have to know more about energy medicine itself, its roots and its basic concepts. In this chapter we will return to the beginnings of science and medicine, to a time when they were one. We shall see how the path to the present was not a straight road of gradual enlightenment, but a road with many wrong turns. Many new ideas emerged and flourished, only to become dogma themselves, used as a defense against more new ideas. In retrospect, we can see that a great cycle has occurred in which we have finally returned to some of the concepts that formed the very beginnings of science and medicine.

THE ORIGINS OF ENERGY MEDICINE

Medical philosophy reflects the prevailing view of society toward basic philosophical questions: What is life? Who am I? What is the relation of living things to the nonliving world?

We now tend to look back with condescending amusement at the medicine man shaking his rattle, thinking that we are superior because we are the beneficiaries of great advancements in scientific knowledge and technology. But are we really so much better off? The present evolutionary philosophy holds that we are simply statistical accidents based upon the coming together of the right chemicals under the right conditions. By this way of thinking, random-chance evolution has resulted in the human being, and we are no more than our antecedents were—chemical machines. If we *are* mere machines, then so are our doctors, and modern medicine is simply the repair of machines by other machines. I am not so sure that this scenario is superior to that of the medicine man shaking his rattle.

It is certainly true that we have passed through a series of stages in the disease patterns that beset humanity. While we were almost totally at the mercy of epidemics of infectious diseases in the past, we now have public-health measures and medications that have almost completely eliminated those diseases. However, we seem to have merely traded one set of problems for another. We may now live longer, but we are unhappy with our "quality of life," and we are, to a growing extent, experiencing other problems during our lifetimes. The incidence of cancer, degenerative diseases, and birth defects is rising; new diseases, such as AIDS, have appeared. We seem to be almost on the brink of disaster, with rapidly increasing numbers of sick people and an overused medical-care system that appears unable to cure, care, or cope. Important questions are being raised that threaten the fabric of dogma that the life sciences have been weaving since the scientific revolution of the Renaissance. This fabric, which once seemed "of whole cloth and seamless," has been shown to be incomplete, containing many misdirected stitches and prone to erroneous interpretations.

"Primitive" Medicine

Long before written history, our ancestors lived in a world full of mysterious forces that governed their lives—the cycles of the sun and the seasons, lightning and fire, wind, drought, and storm. Their own bodies were equally full of unknown energies and potentials: life and death, sickness and healing, birth and aging. Humanity, blessed (or cursed) with a questioning intelligence, has an overwhelming curiosity and a need to explain its place in the "scheme of things." The body of beliefs that arose out of this quest was at first one single system encompassing what we now call religion, philosophy, and medicine.

Peoples widely scattered across the globe shared many common aspects of this early belief system. Primarily, they believed in two realities—the world that they saw around them, and the unseen spirit world in which dwelled the forces that energized the world of nature and people. Life was part of the spiderweb of the universe, with everything interlocked and interrelated by spirit or energy. The Earth was the mother that sustained life, and all that existed had been created by some supreme being. All life was endowed with a special energy, a "life force," that vitalized it.

This life energy was one with the great universal forces of the second reality. Disease was the result of otherworldly forces acting upon the patient, and Death was the transfer of the life energy out of the body into the spirit world. Floods, earthquakes, droughts, famines, disease, death, and birth were seen as the results of deliberate actions by the spirits, who were either displeased with or approving of the actions of humanity. People were at the mercy not only of the natural forces of this environment, but also of the mysterious world of this other reality. Often, the life force was believed to have a dualistic nature, and could become imbalanced under the influence of the external forces—resulting in disease.

These internal and external energies together made up the separate reality with which the shaman-healer established contact during an altered state of consciousness (reached through dreams, severe physical or mental stress, meditation, a "spirit quest," or the use of mind-altering substances). Once the healer had connected with the other reality, he would be able to diagnose and treat his patient by influencing either the external forces of the spirit world or the internal forces within the equally mysterious body of the patient. The cure was produced by balancing the dualistic forces within the patient, transferring to the patient forces from the spirit world, or giving the patient the healer's own life force.

These basic concepts expanded as societies evolved and people had time to examine their environments in more detail. In the process, specific forces were discovered in nature that, while equally mysterious as those of the gods and spirits, were *controllable* by the shaman-healer. Because all living things possessed this life force or spirit, even insignificant-appearing herbs could have effects on the human body via their own "spirit properties." Over the millennia, this idea resulted in a primitive, but extensive, pharmacopoeia.

It seems reasonable to speculate that the mobility possessed by

animals and humans was considered a special manifestation of the
life force. The discovery of lodestone, a natural magnet composed
of the natural magnetic material magnetite, is lost in prehistory.
When lodestones were found to be able to move by themselves, they
were believed to have a particularly potent life force, and this mystical power was thought to be able to influence the human life force.
Static electricity, which also produces a "moving force" and is easily
produced by rubbing natural amber with fur, must have seemed as
mystical as the lodestone.

These discoveries, made long before written history in many
societies, were among the most significant events in prehistoric
times. They represented the beginnings of the exploration of the
environment and the dawn of science. Knowledge of the action of
herbs ultimately led to chemistry, and the lodestone and static electricity were the basis for the development of modern physics. These
discoveries were the keys to the beginnings of the science of medicine and the science of life.

The First Medical "Textbooks"

Certainly, by the advent of written history, medicine
had already evolved into a complex belief system centering on the
life force and body energies. Treatments for influencing these energies involved magic, herbs, and the natural forces of magnetism and
electricity. By all criteria, the system of medicine we call "primitive"
was not only well developed and sophisticated, but was actually a
kind of energy medicine. It left behind a legacy of concepts and
techniques based on a belief in the existence of a "life force" that
could be influenced in a number of fashions. These concepts and
techniques were incorporated into the first written medical texts.

To date, the oldest known medical document is the Chinese
Yellow Emperor's Book of Internal Medicine, attributed to
Houang-Ti and thought to have been written sometime around 2000
B.C. It presents the concept of a bodily energy called *chi,* which
worked through the balance of two other opposing forces in the
body, *yang* and *yin.* Disease was thought to be produced when
these forces became unbalanced, and two specific techniques for
restoring balance—acupuncture and moxibustion—were described.

Acupuncture consisted of the insertion of very fine needles into
specific energy points on well-defined lines, or "meridians," of energy flow on the surface of the body. Included with acupuncture was

the practice of placing lodestones over the same energy points, although this was apparently considered less effective than needle insertion. The other technique, moxibustion, consisted of burning small amounts of "moxa" (a small tuft of soft, combustible material) either over the same points used in acupuncture or over the areas of painful irritation.

Both techniques were viewed as influencing an internal energy system by introducing external energy. This external energy was *electrical*, via the insertion of metallic needles; *magnetic*, using the lodestone; or *heat*, in the moxibustion technique. The origin of these treatments and their basic concepts must have antedated the *Yellow Emperor's* text by at least several thousand years, attesting to the sophistication of thought among preliterate peoples.

An Egyptian document, the *Kahun Papyrus*, dates back to approximately 2000 B.C. It relates the use of specific herbal remedies, along with prayers for intercession by the gods. While the *Kahun Papyrus* does not specifically say so, it is known that the Egyptians also used the properties of the lodestone in therapy (in fact, it is widely claimed that Cleopatra wore a lodestone on her forehead to prevent aging).

The Vedas, the ancient religious scriptures of the Hindus, also date back to about 2000 B.C. These dealt with the treatment of many diseases using *siktavati* or *ashmana*, which can be translated as "instruments of stone" and may refer to the use of lodestones in therapy. Finally, it is well known that Tibetan monks used bar magnets in a highly specific fashion to influence the minds of novice monks during their training. It would seem reasonable to conclude that this practice was based on a much older technique involving the use of lodestones.

Certainly, by the beginning of written history, several Eastern civilizations were practicing a kind of energy medicine in which the forces of electricity and magnetism were employed by physicians to influence internal energy systems of the body.

MEDICINE IN GREECE: THE BEGINNINGS OF WESTERN MEDICINE

Western medicine is usually believed to have begun around 500 B.C. in ancient Greece, with the writings of Hippocrates. However, even in those days "technology transfer" occurred, and the medical concepts of Chinese, Indo-Tibetan, and Mediterranean peoples had already found their way into Greek culture. Some 150 years before Hippocrates, Thales of Miletus—often considered the father of European philosophy—laid the groundwork for modern physics and biology. He "discovered" the lodestone and static electricity (from amber, called in Greek *elektron*). He proposed that living things were animated by a vital spirit and that this spirit was shared by the lodestone and by amber. Thales said, "The magnet has soul because it attracts iron," and "All things are full of gods."

In fact, these concepts were common to the ancient world and were probably learned by Thales during his studies in Egypt. However, his most important contribution was the philosophical idea that there were actual *causes* for all things, and that human beings could discover these causes through the application of reason, logic, and observation. This important concept can be illustrated by the difference between the necromancer's dissection of an animal to determine the intention of the gods, and the philosopher's dissection of an animal to discover its anatomy and to learn how it worked. Thales of Miletus made the first step away from mythology toward the feeble beginnings of science.

Hippocrates incorporated many of Thales' ideas into his philosophy of medicine, but his contributions were far more than a simple codification of preexisting ideas. Hippocrates left an indelible stamp upon all further development of medicine with his prolific writings. While the Hippocratic oath is now being replaced by something less "archaic," I was proud to take it when I graduated from medical school in 1948.

In many ways, Hippocrates can be considered to have been the "ideal" physician. I am certain that even today he would be the type of physician we would all want as our family doctor. He was not arrogant or certain in his beliefs. Of all the quotations attributed to him, I like the following best: "Life is short and the Art long, the occasion fleeting, experience fallacious and judgment difficult." "Art," in this quotation, refers to medicine. If only the modern physician, often so sure that he is right, would adopt some of Hippocrates' humility!

Hippocrates also realized that disease was not a single causal relationship between an external agent and a simple machine, but rather that each disease is the complex product of the agent and the body's reaction to it: "Disease is not an entity, but a fluctuating condition of the patient's body, a battle between the substance of the disease and the natural self-healing tendency of the body." Regrettably, these words of wisdom have been largely forgotten by modern medicine.

While believing that a "vital spirit" was responsible for "life," Hippocrates thought that it acted through four "humors": blood, phlegm, yellow bile, and black bile. Disease was thought to be produced by an imbalance among these humors—a concept very similar to the Chinese *chi*, or life force, which acts through the balance of yang and yin. His treatments for diseases included the use of many natural herbs whose properties were known through preexisting medical knowledge.

In a less well-understood passage, Hippocrates says: "Those diseases that medicines do not cure, iron cures; those which iron cannot cure, fire cures; and those which fire cannot cure are to be reckoned wholly incurable." Clearly, Hippocrates tried herbal remedies first; if these didn't work, he resorted to iron, and then to fire. "Iron" is generally translated as "scalpel," but "fire" has not been satisfactorily translated. Because the natural lodestone was known to attract natural iron and convey its magnetic properties to it, perhaps Hippocrates adopted the ancient use of magnets in therapy. In that event, "fire" is translatable as the equally ancient technique of moxibustion. Given this history, the old common thread of a vital spirit expressing itself through balanced energy flow and alterable by the application of natural forces becomes plainly evident in Hippocrates' writings.

One of Hippocrates' achievements was the idea of the "medical school," wherein prospective physicians could learn their art and craft. He founded many such schools, or *aesculapiae*, throughout the Eastern Mediterranean. Two hundred years after Hippocrates' death, the *aesculapian* at Alexandria, Egypt, produced a remarkable physician and scientist, Erasistratus, who was probably the first man to scientifically dissect the human body. He discarded Hippocrates' theory of humors and linked disease with the abnormalities of internal organs he found by dissection.

Erasistratus properly identified motor and sensory nerves and traced them to the brain, which he believed was the seat of the mind and soul (rather than the heart, as Hippocrates had proposed). He also described the function of the heart as a pump for the blood. While he described the "mechanics" of the body, he was a vitalist who believed that the life force was a subtle vapor he called *pneuma*. In many ways, Erasistratus was far ahead of his time. Had his ideas, which were essentially correct, gained acceptance, medical and biological knowledge would have progressed far more rapidly than it did. Unfortunately, his observations and ideas persisted for only a few hundred years, and were then totally swept away by a graduate from a medical school at Pergamon. That graduate's name, Galen, is well known even now.

Galen was, in most respects, the antithesis of Hippocrates—he was absolutely sure of himself and his beliefs, arrogant, self-serving, and prone to falsehood if it served his purposes. Wise enough not to directly challenge the great Hippocrates, Galen endorsed the concept of the four "humors" but added much additional material derived from his own observations and experiments. Most importantly, he proposed the attractive idea that for every disease there was a single cause and a single cure, which was eagerly adopted by physicians who, then as now, sought authoritarian infallibility. Galen was a prolific writer, and during his lifetime he published a complete "system of medicine," addressing anatomy, physiology, and therapeutics. This became the standard text and ultimately the overwhelming dogma that dominated medicine for the next 1500 years.

Unfortunately, Galen was wrong. His ideas about anatomy were incorrect, and his teachings on physiology were based upon falsified experiments. In his day his concepts were challenged by those physicians who were followers of Erasistratus. Galen responded with what can only be called a deliberate campaign of falsehoods and vilification. He "repeated" the experiments of Erasistratus and found them to be "incorrect." Actually, the opposite was true; Erasistratus was a careful experimenter and observer. However, Galen's dubious scientific integrity was never called into account, and no one bothered to repeat his experiments. While practically all the writings of Erasistratus were destroyed, Galen's writings have been well preserved.

Galen succeeded by providing a comprehensive system of medicine mixed with pseudoscience that provided definite answers for both diseases and treatments. Though largely wrong, it carried a stamp of authority and effectively put a stop to any valid experimentation or questioning for the next 1500 years. Early attempts at logical observation by Erasistratus and the humanism of Hippocrates' "art" were both submerged by the false dogma of Galen. The first wrong turn had been taken. Western civilization entered what historians have (for good reason) called the Dark Ages, in which medicine and science were totally authoritarian and included erroneous concepts of how the body worked.

THE RENAISSANCE: THE BEGINNINGS OF SCIENTIFIC MEDICINE

The emergence of Western civilization from the Dark Ages was primarily the result of one factor, the challenge to authority. In medicine and science, the first challenger was a man who was a strange mix of humanism, mysticism, early scientific logic, and a most abrasive personality. Paracelsus was source of the legend of Dr. Faustus, who sold his soul to the devil in exchange for knowledge. Paracelsus had no respect for authority in any form. At the age of fourteen, he left home to wander across Europe and Asia, studying at many universities and, possibly, never graduating from any. His attitude toward organized learning is best illustrated by his statement that "the universities do not teach all things, so a doctor must seek out old wives, gypsies, sorcerers, wandering tribes, old robbers, and such outlaws and take lessons from them. Knowledge is experience."

Paracelsus detested Galen as an absolute fraud. He once burned Galen's books in front of the university before a "cheering throng of medical students." He stressed the fact that the body can heal itself, while the most that Galen's medicine could do was to delay healing or produce disastrous complications. Paracelsus foreshadowed the discoveries of modern antibiotics by correctly showing that mercury could cure syphilis. He accurately described the cause of thyroid goiter, and he provided the basis for homeopathy by claiming that diseases could be cured by minute doses of "similars"—chemicals that produced the same symptoms.

In the first environmental medicine study, Paracelsus correctly ascribed silicosis in miners to inhalation of materials from the mine, rather than to punishment by mountain spirits. His experiments with herbal remedies and alchemy set the stage for the future growth of chemistry. And he made extensive use of the lodestone in his treatments.

This remarkable man even brushed aside the attacks of organized medicine and science. His fame was great and his lectures (to which all citizens were invited) were filled to overflowing. His writings were remarkably influential, particularly his major work, the

Great Surgery Book. Yet, he remained practically penniless throughout his lifetime.

Many of Paracelsus' beliefs and ideas appear mystical and, to some present-day reviewers, downright crazy. Still, there is no escaping the fact that many were actually profound insights into the nature of disease and physiology that had no precedents at the time. From what secret sources did he derive these creative insights and remarkable concepts? Was Paracelsus a natural "extrasensory perceptionist" who, operating by intuition alone, could make remarkable quantum leaps in understanding?

One of his most startling statements was this:

> To think is to act on the plane of thought, and if the thought is intense enough, it may produce an effect on the physical plane. It is very fortunate that few persons possess the power to make it act on the physical plane because there are few persons who never have any evil thoughts.

Evidently, Paracelsus believed that thoughts had a physical reality and could, "if intense enough," produce an action "on the physical plane." These concepts obviously relate directly to the present-day ideas of extrasensory perception, psychokinesis, and parapsychology. While never explicitly stating so, this passage may also be interpreted to mean that Paracelsus also recognized the power of thought, or conviction, to heal.

Paracelsus was a vitalist who believed intensely in a life force he termed *archaeus,* which could be influenced by the mysterious action of the magnet. He managed to combine this mysticism with a remarkable insight into future biological knowledge. He said:

> The power to see does not come from the eye, the power to hear does not come from the ear, nor the power to feel from the nerves; but it is the spirit of man that sees through the eye, hears with the ear, and feels by means of the nerves. Wisdom and reason and thought are not contained in the brain, but belong to the invisible and universal spirit which feels through the heart and thinks by means of the brain.

The relationship of this concept to the modern concept of the duality of mind and brain is obvious.

Going even further, Paracelsus viewed the body as a whole

organism, composed of many parts. Each part interacted with the others and was inseparable from the whole, which was greater than the sum of the parts: "Even the ignorant knows that man has a heart and lungs, a brain and a liver and stomack [sic]; but he thinks that each of these organs are separate and independent things that have nothing to do with each other." Several centuries before the rise of reductionism, Paracelsus saw the defect in reductionist philosophy and placed its future proponents among the "ignorant."

Paracelsus built his system of medicine on the legacy of the life force, herbalism, and the therapeutic application of natural forces, which can be traced back through the Greeks to prehistoric man. Like Erasistratus, he trusted logical observation and experimentation; unlike Erasistratus, he succeeded in influencing the course of history. His legacy, derived from a visionary power, gave rise to all that we now call science.

In a remarkably prescient statement Paracelsus wrote that "the human body is vapour materialized by sunshine mixed with the life of the stars." Today, physicists believe that the elements of the human body were originally generated in supernovas, the great thermonuclear explosions of stars.

Paracelsus died under mysterious circumstances at age forty-eight, leaving his few possessions to the poor and his remaining manuscripts to a simple barber-surgeon.

THE SCIENTIFIC REVOLUTION UNDER WAY

Two years after Paracelsus' death, Andreas Versalius, a military surgeon, published the first really accurate anatomical text, *De humani corporus fabrica* (The fabric of the human body). This work finally, and completely, dispelled the dogma of Galen's infallibility. The age of science and reason had begun. People began to learn about the science of living things (biology) and the science of nonliving things (physics). The mysterious natural forces of electricity and magnetism gradually began to be understood.

A few great scientists contributed the basic concepts and provided the foundation upon which the rest of science built its edifices. The first of these was William Gilbert, physician to Queen Elizabeth I and the first true scientist whose interest lay not only in medicine but also in the forces of electricity and magnetism. His publication in 1600 of *De magnete* (The magnet) clearly delineated electricity and magnetism as separate forces, established the rules of action for each force, and described the Earth as a large magnet. No longer was it believed that the compass needle points due north because of mysterious rays from the North Star!

Gilbert's most important contribution was, in the tradition of Hippocrates, Erasistratus, and Paracelsus, a plea for "trustworthy experiments and demonstrated arguments" to replace "the probable guesses and opinions of the ordinary professors of philosophy." This plea was later expanded and codified by Francis Bacon in *The Scientific Method*.

During the 1600s, several means of storing electrical "fluid" were discovered, and better methods of generating static electricity were devised. However, knowledge of electricity was limited to static electricity—the same type as that produced by rubbing amber with fur, or by walking across a rug. Knowledge of how living things actually worked also advanced at this time, particularly with the discovery that nerves transmit sensory information and cause muscle contraction. The brain became firmly identified as the seat of thinking and memory. With this knowledge came increasing controversy between the *mechanists*, who viewed living organisms as complex machines that are completely understandable by means of physical principles, and the *vitalists*, who believed in the mysterious, unknowable life force.

However, even among the mechanists there was apparent reluctance to completely exclude the mystery. René Descartes, the main proponent of the mechanistic model, still postulated a "soul," which he conveniently located in the pineal gland (a curious, pine-cone-shaped structure located in the center of the head).

Mesmer, under the influence of Paracelsus' teachings, proposed that living things generated universal forces that they could transmit to others through "animal magnetism." He began treating a variety of ills using magnetic therapy; because he was remarkably successful, he incurred the wrath of the medical establishment. The orthodox physicians claimed that he was practicing magic, and in 1784 King Louis XVI was forced to appoint a commission to investigate him. The commission's report was "unfavorable," ascribing Mesmer's successful results to simple suggestion. The only remaining legacy of his work is the term *mesmerism*, a synonym for hypnosis.

Hahnemann, building on Paracelsus' "Law of Similars," constructed a complex system of medicine known as "Homeopathy."

This system was based upon the administration of minute doses of the "essences" of substances that produced symptoms similar to those from which the patient was suffering. Hahnemann postulated that these essences reacted with the energetic vital spirit of the body in a manner similar to that of the lodestone, a treatment method he also advocated.

Throughout this period of scientific excitement, the argument between the mechanists and the vitalists heated up, with the vitalists eagerly embracing electricity as the scientific life force. In doing so, however, they were putting all their eggs in one basket, for if electricity were ever totally excluded from life processes, they would have lost the battle.

In the late 1700s another remarkable physician, Luigi Galvani, stepped into this controversy. While he was a thoroughly humanistic physician in the Hippocratic tradition, Galvani was also caught up in the fervor of the scientific experimentation of the time. He established his own laboratory, complete with the latest instruments for generating sparks of static electricity by friction. He was searching for proof of the electrical nature of the life force, and he believed that he had found it when he observed muscles contracting when they were connected to the spinal cord with metallic wires. Galvani termed this "animal electricity," and he postulated that this electricity was produced by the living body itself. For some reason, he brushed aside the fact that this effect could be produced only when two wires composed of different metals were used.

Allesandro Volta, a physicist and a colleague of Galvani's, at first supported Galvani's observations. However, he then discovered that the electricity was actually produced by the junction between the two dissimilar metals, and that it was quite different from the single spark of static electricity. What Galvani had actually found was *direct current*, or continuously flowing electricity—a discovery that has shaped the world ever since. Volta's "pile" of dissimilar metals was the beginning of the storage battery and the possibility of continuous generation of large amounts of electricity.

Galvani never publicly responded to Volta's critique. This was unfortunate, because he actually *had* shown "animal electricity" flowing from injured tissue; muscle contraction could be caused without the use of wires, simply by bringing the muscle into contact with the cut end of the spinal cord itself. This later became known as the "current of injury," which is an electrical current found in any injured tissue. But by that time Galvani had been so discredited that the idea of the current of injury was relegated to the status of an unimportant curiosity.

Galvani, like Paracelsus, was far ahead of his time. He observed and reported the transmission of electrical force across space, when a spark produced by his electrostatic machine caused the contraction of a muscle held with metallic forceps by an assistant across the room. This important principle remained "undiscovered" until Hertz's experiments 100 years later. Galvani even searched for variations in atmospheric electricity using antenna wires! Had he defended himself more vigorously against Volta's attacks and continued his observations, the path of science might have been very different.

Fifty years after Galvani's experiments, Emil DuBois-Reymond discovered that the passage of the nerve impulse could be detected electrically. Believing that he had "identified the nervous principle with electricity," he postulated that the nerve impulse was the passage of an amount of electrical "fluid" down the nerve fiber. By then, the total importance of the nervous system for all life functions was well established. The vitalists celebrated the fact that electricity had again become the life force, acting through the brain and nerves. But this happy state did not last long. Within a year, Hermann von Helmholtz had electrically measured the speed of the nerve impulse and found it to be very much slower than that of electricity in a wire. He concluded that while the passage of the nerve impulse could be *measured* electrically, it was not actually the passage of a mass of electrical particles.

In 1871, unhappy with this new status of electrical force, Julius Bernstein proposed an alternate chemical explanation for the nerve impulse. He believed that the ions (charged atoms of sodium, potassium, or chloride) inside the nerve cell differed from the outside tissue fluid, and that this difference resulted in the nerve-cell membrane's being electrically charged, or "polarized." In Bernstein's view, the nerve impulse was a breakdown in this polarization that traveled down the nerve fiber, accompanied by the movement of these ions across the membrane. This, he believed, was what DuBois-Reymond had measured. The "Bernstein hypothesis" was eagerly accepted and has since been shown to be essentially correct, not only for nerve cells but for all cells of the body.

The success of the Bernstein hypothesis resulted in the dogmatic view that this type of electrical activity is the *only* type permitted in the body. In this view, direct electrical currents cannot exist either within the cell or outside it, and externally generated electrical currents (provided these are below levels causing shock or heat) cannot have any biological effect. The vitalists, banking on the mysterious electrical force, appeared to have lost the battle. While there is now no doubt that Bernstein was correct and that membrane polarization is the basis for the conduction of the nerve impulse, it did not necessarily follow that the nerve impulse is the only method of data transmission in the nerve, or that such membrane polarization is the only way that electricity can work in the body. Orthodox science, however, discarded such ideas as vitalism.

By this time, the anatomists with their microscopes had found that the nerve did not actually contact the muscle, and that a space existed between it and the muscle, which came to be called the "synaptic gap." The vitalists were pushed into this small space and forced to postulate that the passage of the nerve impulse across the synaptic gap was electrical.

The argument remained unsettled until 1921, when physiologist Otto Lowei proved by experiment that transmission of the nerve impulse across the synaptic gap is also chemical. (Because Lowei was a research professor at the medical school I attended, it was my misfortune—and that of my classmates—to have to repeat his experiment in the physiology laboratory. I can say, unequivocally, that it worked!) As a result of Lowei's experiment, all traces of electricity and magnetism were firmly excluded from any functional relationship with living things. Vitalism was finally dead.

However, this triumph of science was still surrounded by a mystery. Lowei was still considered a professor at the university even though he was elderly and retired, and he often visited the physiology laboratory, telling the students about the strange events that had surrounded his successful experiment. He had, he said, been wrestling for some time with the problem of how to do the experiment. One night, he had a dream in which the exact way to do the experiment was revealed to him. Unfortunately, when he awoke he couldn't remember any details! The next night he had the same dream, but this time he remembered it all upon awakening. He immediately went to his lab, and in a few hours had successfully demonstrated the chemical nature of synaptic transmission. Lowei's dream ultimately led to his receiving the Nobel Prize in 1936. In his visits to the physiology lab, he would caution us that we did not know *everything*, that some mysteries still remained.

By the turn of the century, the idea that medicine should be based totally on science had become popular. As a result, scientific medicine—based upon the chemical-mechanistic model—was firmly established. Its conclusive proof of efficacy came in 1909, when Paul Ehrlich discovered that the cure for syphilis lay in a specific arsenic compound. Ehrlich called this a "magic bullet," a chemical specifically designed to seek out and destroy the bacterium that was its target. He further predicted that for the rest of the twentieth century, medicine would be characterized by the discovery of similar specific "magic bullets" for all diseases.

As Ehrlich predicted, this concept has dominated modern medicine. The allure of the simplistic and infallible cure is as strong today as it was in Galen's time.

While medical scientists had "conclusively" shown that neither electricity nor magnetism played any role in living things, the physicists and engineers had not been idle. By the 1920s, they believed they had learned all there was to learn about these two forces. We were already enjoying the luxury of electric lights, courtesy of Thomas Edison, and listening to the radio. We were able to generate, transmit, and use these forces, and we understood their characteristics. A new world, based on science and technology, was dawning.

At that time, it appeared to be firmly established that the only way an electrical current administered to the body could have any effect was if its strength were high enough to produce shock or burns. Electrical force below this level simply could not have any effect. Exposure to an electromagnetic field was even more biologically tenuous. If the field was steady-state (direct current, or unvarying with time)—such as from a permanent magnet—it could exert a moving force only on structures or particles within the body that were themselves magnetic. Since no such magnetic material existed in the body, there could be no effect from DC magnetic fields.

Furthermore, while time-varying (pulsing, or AC) magnetic fields could theoretically induce electrical currents within the conducting solutions of the body, these currents would be very much smaller than those required to produce shock or heat. So, again, there could be no effect. As for living organisms producing external magnetic fields, as Mesmer believed, the question was too ludicrous even to consider.

The physicists, biologists, and physicians were absolutely certain that the life force simply did not exist, and that all living things were simply chemical machines. They *knew* that life was simply the result of a chance, random event between chemicals, and that it would occur in a similar fashion wherever the circumstances were right. They *knew* that for each disease there was a single cause and a single therapy, and that the only valid therapy was either surgical or chemical. Finally, they *knew* that the living organism was simply a collection of structures, which worked chemically and were integrated by means of the central nervous system, with no involvement of electricity or magnetism.

Life had been reduced to chemical machinery. The second wrong turn had been taken. As we shall see, the new scientific revolution has shown that the whole of the body is more than the sum of its parts, that the ability of living things to heal themselves is far greater than the mechanists thought, and that electricity and magnetism are at the very basis of life.

THE TRIUMPH OF TECHNOLOGY

Ehrlich's belief that science would ultimately provide us with "magic bullets" that would cure each and every disease seemed to come true during and following the technological explosion of World War II. The antibiotic penicillin revolutionized the practice of medicine and, more importantly, provided the hope that other chemical agents might be found that could selectively produce desired effects on cancer and other degenerative diseases.

The discovery of DNA as the basis of heredity strengthened the Darwinian concept of random evolution, and it diminished the stature of human beings to that of simple machines controlled by the composition of their DNA base pairs. Advances in the understanding of the body's chemical processes and in surgical techniques, such as the use of artificial organs and living transplants, have nearly brought to fruition the dream of Dr. Frankenstein—the cobbling together of a human being from separate parts. We are now preparing to alter our genetic material to produce a "better" person. The idea that artificial organs are the same as, or even better than, those we were born with is common in the popular mind.

In many ways, modern medicine has gone beyond scientific medicine to a new phase, that of technological medicine, based upon the applications of this technological revolution. At the same time, technology has become dominant in society and in our lives. Electromagnetism has become the "dynamo" of our civilization, and through its use for power and communications we have succeeded in changing our environment more radically than ever before. Since science had totally excluded electromagnetic forces from life, we eagerly accepted these remarkable "advances" without even questioning their possible biological effects. The third wrong turn had been taken.

This technological revolution, now forty years old, has begun to show its defects. The medicine it has produced is increasingly complex, expensive, and inadequate. We have failed to find magic bullets for anything other than infectious diseases, and we are facing a spectrum of new diseases against which technological medicine appears helpless. The mechanistic view of life has gradually been shown to be unable to provide satisfactory explanations for the basic functions of living things. Its proponents have confused the machinery of life with life itself, and in the process have managed to learn more and more about the machinery but less and less about life.

Two revolutions are now afoot: First, patients and physicians alike have begun to explore new avenues and approaches for dealing with these mounting problems. Second, scientists—both physicists and biologists—have set about constructing a new paradigm that reintroduces energy back into life. In the next chapters we will explore this new scientific revolution. We'll see how it is bringing about the rehabilitation of some ancient beliefs, and how it is influencing the development of new medical therapies while at the same time revealing the unexpected and disturbing consequences of our unbounded use of electromagnetic forces.

CHAPTER TWO

THE NEW SCIENTIFIC REVOLUTION: THE ELECTRICAL CONNECTION

A science with all the answers and no more questions atrophies into mere technology.

MICHAEL COLLIER,
Introduction to Grand Canyon Geology

O ver the past few decades, scientists have begun to realize that mechanistic concepts penetrate merely the outer layers of life, and that under this superficial cloak lie many unexplained mysteries. Recognition of these defects in the present chemical-mechanistic theory have prompted questioning and experimentation. Gradually, a new scientific concept has emerged, one that has brought energetic systems back into biology and has begun to explain many of these mysteries. In particular, these energetic mechanisms have been shown to be the basis of many of the underlying control systems that regulate the complex chemical mechanisms. As we move along this new pathway, a richer biological and medical philosophy is emerging.

In order to understand these advances and the changes that they will bring about, we will begin with the mysteries of life that the chemical-mechanistic concept has failed to explain, and we will see how the new biology brings us closer to understanding the reality that integrates life, energy, and medicine.

UNDERSTANDING GROWTH AND HEALING

We are all familiar with growth. Our cuts heal, our fractures mend, we watch with love and pride the growth of our children. Physicians knew early on that some wounds did not heal,

27

some fractures failed to mend, and some children grew abnormally. There have been countless attempts to understand this mysterious growth process, but all have failed. The advice given to doctors in training from the time of Hippocrates to the present day has been that nothing can be done to speed up, encourage, or restart a failing growth process. The best one can do is to not get in the way of Mother Nature and not do anything that would impede healing.

The most familiar growth process is the healing of minor wounds. If the cut on your finger is deep enough, you know that you'll be left with a scar after it heals. This results from the most common form of healing in human beings, known as fibrosis. Fibrosis is a relatively simple closing of the wound edges with a fibrous tissue called collagen, made by specialized fibroblast cells.

Technological medicine "explains" this process simply by describing its details. The injury produces bleeding, which forms a clot that seals off the wound. The small blood platelets within the clot release a chemical called platelet-derived growth factor, which activates the DNA within the fibroblast cells in the tissues, causing them to begin making collagen fibers. As these fibers are released into the wound, they contract and gradually draw the wound edges closer together. At the same time, the skin cells at the edges of the cut begin dividing and making new skin cells. These migrate over the fibrous tissue, and the wound heals. In this view, the healing process is strictly "local"—that is, it is completely unrelated to the rest of the body and could even occur in a test tube.

This sounds appropriately scientific and gives the impression that we know everything there is to know about wound healing. However, a number of questions remain: What starts this process? What tells the cells exactly what to make? What stops the process when the wound is healed? Further, this is the *simplest* form of healing; there are others that are far more complex and that are completely related to the rest of the body, and for which technology has no explanation.

For example, we take the growth of any embryo for granted. The fertilized egg grows into the adult organism, be it an earthworm, a shark, a mouse, or a human being. The fertilized egg is one single cell, with little visible organization—yet it becomes a living organism of immense complexity, composed of billions of different kinds of cells. Orthodox science tells us that the original egg contained all of the instructions, coded in its DNA, to produce the different types of cells that make up the adult. The formation of the different mature cell types is accomplished by the "repression" of all of the DNA *except* that which is coded for the specific cell type required. For example, the only DNA that's active in muscle cells is that for "muscle"; the rest of the DNA, while still there, is repressed and inactive. In simple terms, "muscle" DNA causes these cells to make certain types of proteins and intracellular structures that are characteristic of what we call a muscle cell. Hence, the muscle cell has the anatomical and functional characteristics that we call "muscle" because of the action of the "muscle" DNA alone.

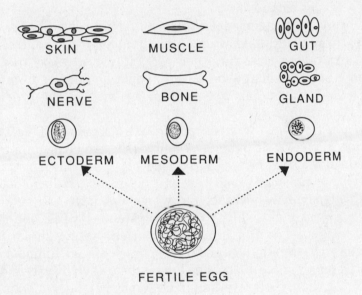

FIGURE 2-1. *The formation of mature cells from the original fertilized egg. The egg contains the DNA instructions to make every one of the different types of cells present in the adult body. Each is produced when all of the instructions but one are gradually repressed. Early in the process, the three main tissue types of the adult (ectoderm, mesoderm, and endoderm) are sequestered; the active DNA for each is restricted to the cell types that each will ultimately form (skin and nerve cells, for example, come from the ectoderm). The final mature cells retain the instructions for all cells; however, only one set of instructions is operating. In theory, then, we could take any mature cell from an adult and activate* all *of the DNA. If the cell then began to divide, it would recapitulate embryonic development and produce a "clone" of the original donor. This, of course, has not yet been done.*

This process is called *differentiation*, and on the surface it appears to explain reasonably well how embryonic growth occurs. However, living things are not just masses of different kinds of cells brought together in a lump. The cells are organized into tissues and the tissues into organs. Most importantly, these are together organized into a discrete entity, the living organism. Mechanistic science tells us nothing of how the intricacies of this organization are produced. It doesn't tell us, for instance, how the DNA "knows" to make muscle here, bone there, and all the remaining organs and structures in their appropriate relationships—much less how it then ties them all together with a brain and a nerve system, so as to create a living organism.

Obviously, what science has done is simply describe the growth process. What is missing is an understanding of the control system that starts this process and then regulates it so as to produce the desired result. Nothing—be it the simple healing of a cut or the fertilization of a human egg—will occur without the operation of some such system.

When I began my work on growth processes, I had already developed a theory of biological control systems based on concepts derived from electronics, physics, and biology. I theorized that the first living organisms, whatever their nature, *must* have been capable of self-repair; otherwise, life would not have succeeded. Self-repair requires a closed-loop control system—that is, one in which a certain signal indicates injury and causes another signal to effect repair. As the repair proceeds, the injury signal diminishes, and when the repair is complete the signal stops. Such a system does not require consciousness or intelligence, so it may be simple in nature but indistinguishable from life itself.

It also seemed to me to be realistic to assume that once such a system had originated, it would continue to serve the same purpose as organisms increased in complexity. Going up the evolutionary ladder from simple organisms to the human being, complexity would increase in anatomical structures and chemical reactions, but underlying everything would be the ancient control systems. Therefore, while much of my work involved growth and regeneration in unusual animals such as salamanders and frogs, I theorized that what I was studying was essentially the same for all life forms, and that my findings regarding these animals would therefore be relevant to human beings as well. This chapter will present findings from my laboratory and from the labs of many other scientists, setting the stage for the discussion of the applications of these findings in the sphere of human health.

Over the past forty years, science has developed a complete theory of control systems. This theory has led to our present-day computers, and lately it has begun to be applied to living systems as well. We use a wide variety of electromechanical control systems to regulate many of our everyday activities, from manufacturing goods to regulating the temperatures of our homes. A common characteristic of such control systems is the regulation of a powerful process by a signal of very low power. For example, in the common home oil-heating system, an electrical signal of only 12 volts and very low current senses the room temperature and controls the operation of the burner unit, which operates at 110 (or 240) volts and several amperes of current. The end result of this action is that the temperature of the room is held fairly constant at the desired level.

The new scientific revolution against the chemical-mechanistic concept resulted from applications of new physical concepts of electricity and control-system theory to the problems of growth and healing. It began when investigations found that there were minute

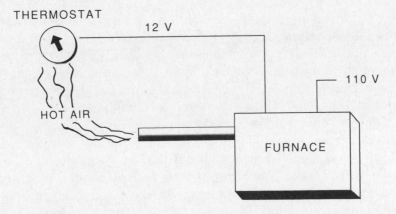

FIGURE 2-2. *The common home-heating system is an example of a negative-feedback control system. The thermostat senses a drop in room temperature and sends a low-voltage signal to the furnace. This low power signal turns on the oil burner and air blower, which are high-power units. The furnace makes hot air, which is blown into the room; as the temperature rises, it reduces and cuts off the signal from the thermostat. The property controlled is the room temperature. In this case, the low energy signal to and from the thermostat regulates a much higher energy system, the furnace.*

electrical currents flowing in organisms, and that these appeared to be the control signals that started and regulated growth and healing. This revolution does not replace the chemical machinery that so fascinated the mechanists; rather, it simply shows us what the switch is that turns the machinery on.

Theoretically, biological control systems should be similar to such inorganic systems as a computerized home-heating system. In general, a biological system could be presumed to have a mechanism for sensing an event (for example, an injury), and then transmitting this information to the central processing unit (in this case, the brain). In the brain, an automatic "decision" would be made to initiate repair of the wound, and a stimulating or "correcting" signal would then be transmitted from the brain to the site of injury. As repair took place, the injury signal would be diminished, causing the stimulating signal to also diminish. System activity would cease when the repair was completed. This is a simple negative-feedback control-system loop. I postulated that such systems in living organisms would very likely be electrical in nature.

THE MYSTERY OF REGENERATION

Regeneration is one of the healing processes that the chemical-mechanistic philosophy has failed to explain. It is the ability not just to heal, but to completely *replace* missing body parts. Many "lower" animals have this ability. It is common knowledge, for example, that half an earthworm will become a whole earthworm if given a little time. To many scientists, this does not seem to be much of a trick; earthworms are pretty simple things. This view does a great disservice to earthworms, which in actuality are fascinating and complex animals that exhibit astonishing behavior!

The salamander is another animal with an absolutely startling ability to regenerate missing parts. On the evolutionary scale, the salamander is an amphibian that stands a notch below the frog. From our lofty perch, amphibians may seem pretty far down the ladder. However, we are both actually in the same boat, since both human beings and amphibians are classified as vertebrates (that is, animals with a backbone and bony skeleton). The salamander has the distinction of representing the basic vertebrate from which the rest of the "higher" animals, humans included, have been derived. The salamander's anatomy is a duplicate of ours (or vice-versa!). While our coccyx (a few tiny bones at the base of the spine) is a poor vestige of the salamander's tail, the rest of our bodies are remarkably similar. The salamander's foreleg has the same bones, muscles, blood vessels, and nerves, in the same arrangement, as a person's arm. The brain and the arrangement of nerves throughout its body are basically the same as ours, except that the thinking area of our brains is greatly expanded. The salamander's heart has three chambers; ours has four. In short, the anatomical complexity and arrangement of this animal are remarkably similar to ours.

The salamander, so like us in its body plan, is capable of regrowing in full detail a foreleg, hind leg, eye, ear, up to one-third of its brain, almost all of its digestive tract, and as much as one-half of its heart. Its growth-control systems are so effective that they also apparently prevent it from getting cancer. In theory, then, the salamander is capable of immortality—unless it is eaten.

In exchange for increasing complexity of structure, we have lost most of this remarkable healing ability. We speak of regenerating severed nerve fibers in the hand, but this is simply regrowth of the fibers from the undamaged cells in our brains or spinal cords. In actuality, the only true regenerative growth we still possess is in the healing of bone fractures. It is no wonder that since the regenerative abilities of the salamander were first discovered by Lazzaro Spallanzani in 1768, scientists and physicians have speculated on how this remarkable property could be restored to us.

The actual process of regeneration flies in the face of some very basic tenets of the chemical-mechanistic dogma. The concept of the whole being simply the sum of the parts led to the reductionist idea that each organ could be removed, placed on the lab bench, and, if properly cared for, could function as well as it did when it was part of the whole organism. The organism was simply a collection of individual parts bolted together. In this view, healing processes were localized phenomena, having no relationship to the rest of the total organism and resulting simply from local factors. This view works fine for human beings, because we don't have any growth processes that challenge this localized view.

Observation of the salamander, on the other hand, completely refutes this idea. Cut off the right foreleg of a salamander, and it will grow a new one, complete in anatomical detail and fully related to the rest of the animal. The salamander does not make any mistakes in its regenerative growth; it doesn't grow a left foreleg, or an upside-down leg, or one that is unconnected to the rest of the animal's body by the appropriate nerve, muscle, and bone connections. The end result is a fully functional, accurate anatomical copy of the original.

Obviously, the process of regeneration in the salamander *must* be completely related to the entire remainder of the organism by some energetic mechanism that encompasses and organizes the total organism in a fashion that cannot be explained by the chemical paradigm. To watch this miracle of limb regeneration occur is the most convincing evidence I know of for the awesome and still unknown power of life.

The details of regenerative growth are even more astonishing in what they reveal of the capabilities of living cells. The first thing that happens after a salamander's limb is amputated is a rapid growth of the outermost cells of the skin across the cut surface. A day or two later, the cut ends of the nerves in the stump grow and make an unusual connection with each skin cell, called the neuroepidermal junction (NEJ). This connection is essential for regeneration, and any technique that prevents its formation will also prevent regeneration.

Shortly after the formation of the NEJ, a mass of primitive cells appears between the cut end of the stump and the NEJ. This mass of embryonic cells is called the blastema, and it is the raw material from which the new limb will grow. We now know that these primitive cells come from the mature cells of bone, muscle, and so on that remained in the stump and somehow returned to an embryonic state. This process of "rewinding the tape" of embryonic growth is called *dedifferentiation*, and it is the key element in bringing about regeneration.

When I began my experiments on the salamander, the concept of dedifferentiation was considered heresy. The scientific thinking of the time required that the process of repressing DNA to make specific types of mature cells be considered irreversible. The reason for this always escaped me, possibly because it would require that DNA could be both repressed and *derepressed* as required, giving living things too great an ability to regulate their own activities.

Everyone who studied regeneration could see that a mass of primitive cells formed at the cut end of a leg. The problem was in explaining how these cells had gotten there within the constraints of the dogmatic commandment that said, "Thou shalt not consider dedifferentiation." Consequently, a lot of time, effort, and money went into experiments that supposedly showed the migration of primitive cells from other parts of the body to the site of injury.

HOW LIMBS REGENERATE

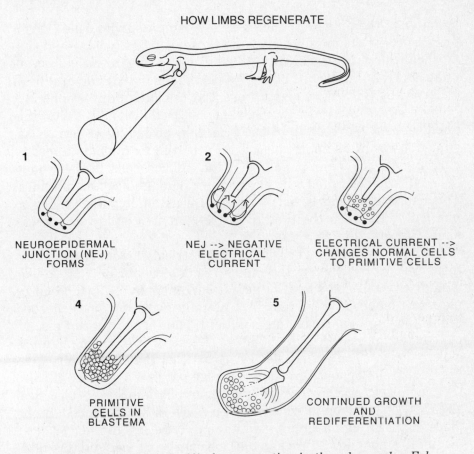

1 NEUROEPIDERMAL JUNCTION (NEJ) FORMS

2 NEJ --> NEGATIVE ELECTRICAL CURRENT

ELECTRICAL CURRENT --> CHANGES NORMAL CELLS TO PRIMITIVE CELLS

4 PRIMITIVE CELLS IN BLASTEMA

5 CONTINUED GROWTH AND REDIFFERENTIATION

FIGURE 2-3. *The sequence of limb regeneration in the salamander. Following the amputation, the skin grows over the end of the stump and the nerves grow into it, making the neuroepidermal junctions (NEJs). After this arrangement is completed, the mature cells of bone, muscle, and so on in the stump dedifferentiate and become a mass of primitive cells, called the blastema. These cells multiply and begin to differentiate into the specific cells needed to make the new arm.*

Despite the fact that no one was ever really successful in showing this, the doctrine that dedifferentiation could not occur remained uncontested. On this fragile basis, the additional dogma was established that human beings failed to regenerate because they did not have such a supply of primitive cells in reserve.

I cannot say what made me do my first experiment on regeneration. For some reason, I decided that it would be fascinating to see whether any differences existed between the electrical currents and voltages at the amputation sites of salamanders (who would go on to regenerate the missing legs) and those of frogs (who would simply heal the cuts by fibrosis). What I was going to measure was known as the "current of injury," a phenomenon that had first been observed by Galvani as being present at the site of any wound in a living animal. As a result of Galvani's discreditation by Volta, this phenomenon had received little scientific attention; everyone assumed it was of no importance. The "scientific" explanation for the current of injury was simply that the injured cells in the wound "leaked" their charged ions across their damaged cell membranes. This explanation was based on the old Bernstein hypothesis, but no one had ever taken the time to *prove* that it was correct. It was simply accepted at face value.

The current of injury is a direct electrical current (DC). If it is measured with a meter, the meter reading would be constant or would vary only slowly. The actual current may be composed of electrons (such as a current in a metallic wire) or ions (charged atoms moving in a conducting solution). The former type of electrical conduction is called "metallic" or "electronic," and the latter is called "ionic." The actual electrical particles that are involved, be they electrons or ions, are called charge carriers.

In all cases, the flow of a DC current implies an electrical circuit, a closed loop in which charge carriers are produced in one place, flow through a circuit, and ultimately return to their place of origin. In a flashlight, for example, the charge carriers (in this case, electrons) are produced by the battery, given by the plus terminal at the front end of the battery to the bottom of the light bulb. They then flow through the filament of the light bulb, producing the light. They leave the bulb from the side of its metallic socket and return to the other end of the battery. The turning on of the switch merely completes the circuit, letting the current flow and producing the light.

When the Bernstein hypothesis was formulated, metallic and ionic conduction were the only two types of electrical current known. Because there were obviously no metal wires in the body, it made good sense to explain everything on the basis of ions moving in a solution. We now know that there is a third type of electrical conduction, semiconduction, in which currents may be either electrons or "holes" (the absence of electrons) moving in a solid material. The discovery of semiconductivity and its industrial application has given us television sets, radios, and tape recorders and has revolutionized our use of electromagnetic energy. We will return to semi-conduction later on.

Regardless of the actual type of conduction, a DC current is produced only when there is a voltage that makes it move. A voltage is produced whenever there is an excess of charge carriers with respect to another point on the same object. Since the current moves in a certain direction, our meter will detect this as either a positive or a negative current.

Not only was this experiment fascinating, but it turned out to have also been a remarkably fortunate choice; the results were dramatic and far more significant than I had expected. I had thought that the salamander would simply have more electricity than the frog. What I found was that the current of injury was *negative* in the salamander and *positive* in the frog—a difference that could not be mistaken.

The experiment completely refuted the theory that the current of injury was merely a by-product of injured cells. First of all, the charged ions in the frog are the same as those in the salamander.

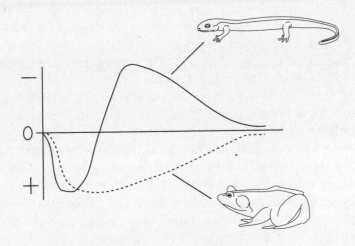

FIGURE 2-4. *The current of injury following limb amputation in the salamander and the frog. Both are immediately positive and remain so for two to three days, at which time the salamander's positive current begins to fall. This is just after the neuroepidermal junction is formed. As the blastema grows, the salamander current becomes highly negative and slowly returns to its original baseline. The frog current of injury remains positive and slowly returns to its original value as fibrotic healing takes place.*

If it were simply a matter of charged ions leaking across damaged cell membranes, the polarity of the wounds in both animals would have to be the same. Second, the current of injury in both animals could be measured for the entire time that healing was taking place (for example, about four weeks during the summer months). It was inconceivable that damaged cells did not either recover or die in, at most, a few days. The electrical currents I measured were active agents, deliberately produced and directly related to the type of growth process the animal used to heal its wound.

The control system that started, regulated, and stopped healing was *electrical*. In the case of regeneration, the electrical current was negative in polarity; in the case of simple scarification healing, it was positive in polarity. While these findings could not be extrapolated to other animals, they gave us, for the first time, firm evidence that electricity was a controlling factor in healing processes. They also raised the possibility that the system that organized the total body was electrical in nature and evolutionarily ancient.

Because it put electromagnetic energy back into biology as a controlling factor, this simple experiment is often referred to as the beginning of the new scientific revolution. The results of the experiment indicated that electricity was one basic element of a control system that regulated regeneration, as opposed to simple scarification. Much more needed to be learned: What was the nature of the electrical current? Where was it coming from? Was it part of a complete control system, with inputs to a master control and outputs that regulated growth?

It appeared possible that there was some relationship between the negative current of injury in a regenerating area and the neuro-epidermal junctions that appeared at the same time that the current began to go in the negative direction. The NEJ is an oddity. It appears only in regenerating wounds and, as noted earlier, is absolutely essential for the regenerative process.

Nerve fibers normally end in special sense organs, such as touch or pressure receptors, or in excitable tissues, such as muscle. They *never* terminate in simple skin cells, where they could have no action. Both nerve and skin are derived from the same basic cell line—the ectoderm—and it is not inconceivable that some peculiar electrical relationship exists between skin and nerve. Since the DC electrical control system had to be evolutionarily ancient, was the NEJ a specific organ, developed very early in evolution, whose job it was to produce the negative DC currents? If so, the next questions to ask were, *How* did it produce DC electrical currents, and what kind were they?

✿

In 1941, Dr. Albert Szent-Gyorgyi proposed that the new method of electrical conduction, semiconduction, played a role in living cells. It seemed to me that this was the appropriate time to look into that possibility.

Semiconductivity is a property of specific materials that have a crystal-like structure—that is, their atoms are arranged in a regular, latticelike fashion. If an atom in the lattice has an extra electron, that electron is free to move through the rest of the lattice, hopping from one atom to the next. Likewise, if an atom lacks an electron, there is then a "hole" in the lattice that can move in the same fashion. The currents in semiconductors are always very small, and only very small voltages are needed to make a current flow. What Szent-Gyorgyi was referring to was the fact that materials, such as proteins, that make up the cells have such organized structures. At the time he made this suggestion, the electron microscope had not yet been invented, so he did not know the true extent to which cells and intracellular structures are organized. Had he known, he would likely have been more convinced of the correctness of his theory. Semiconductivity should certainly have been expected to exist in living things because of their astonishing complexity of organization. For my purposes, biological semiconduction offered the advantage of being the basis for the organized electrical circuits I postulated to be the missing control systems. Ionic conduction simply could not do the job.

In a number of experiments, I was able to show that the DC electrical currents I was measuring from a variety of tissues, including nerve fibers, were actually semiconducting. While the salamander's NEJ is too small to test, it probably functions in the same fashion. This biological semiconductivity is still being explored and has great promise for providing increasing insight into life processes.

As a result of the interest stirred up by these experiments, many people began to make electrical measurements of other growth processes. The findings continued to make sense. The electrical polarity of healing human bone fractures was negative, the same as in salamander limb regeneration. All rapidly growing tissues were found to be negative in polarity. Interestingly, cancers in animals or humans always showed the highest negativity. The consequence of this research was the idea that perhaps with the right amount of electricity we would be able to initiate a growth process.

Our initial experiments on bone growth in dogs, done in collaboration with Dr. Andrew Bassett of Columbia University, seemed to substantiate this. Large amounts of bone growth surrounded the negative electrode, while resorption of bone occurred at the positive electrode. Since this was exactly what we had predicted, we began to think that negative electrical polarity stimulates growth, while positive polarity inhibits it. Unfortunately, these ideas, originating from early experiments done in the 1960s, became so firmly accepted that they persist today, despite being shown to be oversimplistic (as will be discussed in a later chapter).

I spent a long time exploring the implications of our findings, and it was a number of years before I returned to the problem of regeneration. What brought me back to it was the question of the electrical current's producing dedifferentiation. It seemed to me that this was the only process that made any sense. If it was forbidden by science theory, so much the better.

Dr. David Murray, chairman of the Department of Orthopedic Surgery at SUNY medical school, and I had been looking at the healing of fractures (something more appropriate for two orthopedic surgeons!). We chose to use frogs because they were cheaper and easier to work with than rats or mice. This turned out to be a fortunate choice, because we found that the frog simply turned the blood cells in the clot that formed at the fracture site into primitive cells, which then went on to become bone cells. (The frog was able to do this because its red blood cells still contain their nuclei, while those of mammals do not.)

This finding immediately suggested that the red cells were being dedifferentiated by the negative electricity at the fracture site. If so, we should be able to put a frog's normal red cells into a culture dish and cause them to dedifferentiate with a similar negative current from a battery. This sounds simple, but in fact it turned out to be a difficult and lengthy experiment. I had to choose an amount of current to use and, not being very smart, I thought I would need "enough" to do the job. My "enough" turned out to be much more than needed, leading us on a long search for the right amount.

What we finally found was astonishing. The frog's red cells could be dedifferentiated by electricity, but only with vanishingly small amounts (measured in *billionths* of amperes!). Even a little more electricity simply did not work. If we had the right amount of electrical current, in less than an hour we could watch one red corpuscle change from a cell filled with hemoglobin containing a small, shriveled-up nucleus, to a cell with no hemoglobin and a large active nucleus. The changes in the red cells produced this way in the culture dish were identical to those we saw in the blood clot at the frog's fracture.

These results were of fundamental importance in bringing about the present scientific revolution. They clearly showed that the activities of living cells could be markedly influenced only by certain levels of extremely small electrical current. This made the entire concept of electrical control systems feasible and it put an end to the dogma that shock or heat were the only possible biological effects of electricity. A follow-up experiment, done by my colleague Dr. Daniel Harrington, showed that the RNA of these cells was changed exactly as required for dedifferentiation. This helped put an end to the doctrine that dedifferentiation was forbidden. A decade of our looking at living organisms and cells in this fashion had firmly restored electromagnetism to an important role in life processes and had shown that the capabilities of life are far greater than mechanistic science had envisioned.

The natural conclusion was that electricity could possibly be used to control and even start human growth processes. The first clinical experiments, done on human bone fractures that had failed to heal, clearly demonstrated that bone healing could be restarted.

For the first time in thousands of years of medical practice, physicians could actually control a growth process by inserting the appropriate energy. Thus, the first fruits of energy medicine were obtained.

Because I like to push into the unknown, I was less interested in making fractures heal than in the prospect of actually making human beings regenerate lost parts. So, with visions of the Nobel Prize in my head, I decided to see whether negative electricity, applied over a long enough time, could make a new leg grow at an amputation site in the rat. It is said that a good research project yields more questions than answers, and this was such a project.

My colleague, Dr. Joseph Spadaro, and I found that extremely small amounts of negative electrical current administered to an

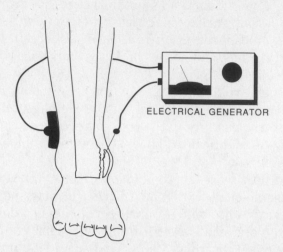

ELECTRICAL GENERATOR

FIGURE 2-5. *A fracture at the ankle had failed to heal. A metal wire electrode was placed directly into the fracture and connected to the negative terminal of a DC generator. The positive terminal was connected to a flat electrode on the skin of the leg opposite the fracture, and the current flowed through the fracture site. In eight weeks the fracture had healed.*

amputation between the shoulder and elbow turned on a regenerative process. An actual blastema formed and grew into all of the missing structures, including bone, muscle, nerve, tendons, and so on, down to the elbow joint itself.

While we could not get the growth to proceed beyond this point, the experiment nevertheless proved two important points. First, electricity was clearly a stimulus to regeneration. Second, and more importantly, the *instructions* required to make a new leg were still retained by mammals. Therefore, the growth control system required for regeneration was present in the rat, and quite likely in human beings as well. However, something was still missing, and it was quite a few years before I learned what it was.

For electricity to produce an effect upon cells, there must be not only the right amount of electricity of the right polarity, but also cells that are sensitive to DC electrical currents. In other words, not all cells are the same. Application of the right polarity and the right amount of electrical current to a sensitive cell will cause it to dedifferentiate. The same electrical factors administered to a cell that is not sensitive will have little or no effect. In the salamander, *all* cells of the body will dedifferentiate if exposed to appropriate negative electrical currents. In the mammal, only certain cells of the bone marrow have this capability. This is what makes us able to heal our fractures.

In our rat experiments, we dedifferentiated these cells and formed a blastema. However, the number of dedifferentiated cells was only enough to form a small blastema, which was able to regrow just the missing part of the upper foreleg, and no more. Since the amount of bone marrow in the human is even smaller than in the rat, there seemed to be no chance that this technique would ever be able to produce regeneration in the human. We had the growth control system that was required for regeneration, but we did not have the numbers of sensitive cells needed to make it work.

We were able to confirm these observations, not only by applying negative electrical currents through electrodes, but also by surgically producing neuroepidermal junctions in rats. We literally stumbled into this finding in the course of an experiment designed to determine what effect a nerve would have on the cells of the bone marrow. In this experiment, which was proposed by James Cullin, a student of mine, we amputated the hind limbs of a number of rats just above the knee joint. We then transplanted the cut end of the main thigh nerve into the bone marrow and brought it out the cut end of the bone. In order to keep it in place, we found that we had to suture it to the skin over the bone end. Two weeks later, we found that we had again started a regenerative process! Just under the skin that had been sutured over the cut end of the bone were new bone, muscle, soft tissue, and nerve fibers. Since we had not applied any electrical current, this was totally unexpected. The only way it could have occurred was if a neuroepidermal junction had somehow formed and produced the necessary negative DC electrical current.

We repeated the experiment, this time with control animals that had the same amputation but no nerve sutured to the skin. After a week, we measured the DC electrical potentials at the amputation sites. The control animals were electrically positive, while those that had the nerve transplanted and sutured to the skin were negative! When we examined the latter with the microscope, we found hundreds of NEJs—and regenerative growth. A small experiment, done out of simple curiosity, had enabled us to make a remarkable discovery: *the neuroepidermal junction is the source of the negative DC electrical current that stimulates regeneration.* But there was still more to be learned.

It seemed apparent that placing the nerve directly over the cut end of the bone enabled the negative DC current to contact the bone-marrow cells, and the blastema that resulted had produced the regeneration we saw. I reasoned that if we did exactly the same experiment but this time sutured the nerve to the skin on the *side* of the leg, away from the cut end of the bone, a neuroepidermal junction should still form and produce the negative DC current. Since there were no bone-marrow cells in that area, however, no growth should occur. When we did this experiment, we found that a junction did, indeed, form and produce the negative DC current on the side of the leg, but, as expected, there was no regenerative growth.

We then had the answer to what caused regenerative growth: The neuroepidermal junctions produce a negative DC current that stimulates sensitive cells in the area to dedifferentiate and form a blastema. Part of the control system that stimulates and controls regenerative growth had been identified. However, we did not know what told the blastema where it was in relation to the rest of the animal, and what it should become.

We had finally solved the riddle of why the salamander can regenerate almost anything and humans can only regenerate bone. While we did find that mammals can regenerate complex structures (and presumably, therefore, do retain at least some of the control system), the number of sensitive cells available to dedifferentiate was found to be insufficient, and only small regenerates were possible. Without the necessary large numbers of these appropriate cells, it appeared that the idea of restoring any major amount of regeneration to the human being was dead, at least for the time being.

DISCOVERING ACUPUNCTURE

In order for a growth control system to start a healing process, it has to receive a signal indicating that an injury has occurred. During the 1960s and early 1970s, it was assumed that this signal was sent to the brain by the sensory nerves that supposedly produced the sensation of pain in the conscious mind. This seemed appropriately simple and in accord with current concepts of neurophysiology. Then President Nixon visited China. When one of the

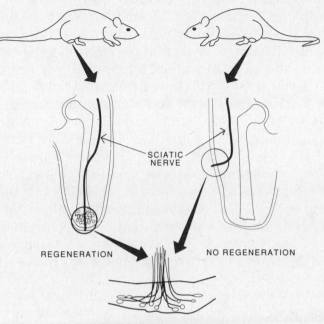

SCIATIC
NERVE

REGENERATION   NO REGENERATION

NEUROEPIDERMAL JUNCTION

FIGURE 2-6. *Surgical production of a neuroepidermal junction in the rat's hind limb. On the left, the sciatic nerve has been transplanted through the bone marrow into the skin. The negative DC potential and a small amount of regeneration are produced by dedifferentiation of the bone-marrow cells. On the right, the sciatic nerve is sutured into the skin on the side of the leg. While the negative DC potential is also produced, no regeneration takes place because there are no cells that will dedifferentiate at that site.*

reporters accompanying him had to be operated on for acute appendicitis with acupuncture anesthesia, a whole new light was thrown on this idea. Acupuncture became a common subject on the evening news.

At that time I recalled that more than a decade before, I had been visited by a colonel from the army surgeon general's office, who had asked me what I knew about acupuncture. I said that I knew only that it was a Chinese technique that involved inserting needles into specific "points" on the skin that were supposed to block pain, but that it made no sense because the points were not located where there were nerves. When I asked him why the army was interested in such an oddity, he replied, "It works, and we think it may be a useful technique for battlefield anesthesia." We talked a bit about how one would go about evaluating it from a scientific point of view, and I suggested that it might be working electrically. He left promising that I would be getting a grant to investigate acupuncture. I never did, but I continued to think about acupuncture. If the *army* thought it worked, there must be something to it!

Nixon's visit must have stimulated the people at the National Institutes of Health, because they sent out the word that they were interested in funding studies on acupuncture. This was my big opportunity, so I proposed to NIH that I and my colleague, Dr. Maria Reichmanis, a biophysicist, look into the possibility that the acupuncture system of points and meridians carried the electrical signals of injury. We would not try to prove that acupuncture did anything clinically; rather, we would simply apply known techniques of electrical measurements to the points and meridians to see whether we could measure anything that made sense. We reasoned that if we could find that there were reproducible and significant electrical parameters associated with the points and meridians, then these would be shown to actually exist. If we found that they did exist, then we could look for whatever anatomical structures produced them and try to discover how they carried information. The NIH funded about twenty project proposals, of which ours was one.

We found that about 25 percent of the acupuncture points on the human forearm did exist, in that they had specific, reproducible, and significant electrical parameters and could be found in all subjects tested. Next, we looked at the meridians that seemed to connect these points. We found that these meridians had the electrical characteristics of transmission lines, while nonmeridian skin did not. We concluded that the acupuncture system was really there, and that it most likely operated electrically. Since clinical studies had shown that relief from pain was the one result that could be reliably produced by acupuncture treatment, this system was quite likely the input route to the brain that transmitted the signal of injury.

To further prove this, I proposed what I thought was an elegant concept. DC currents flowing along a conductor always diminish in intensity the further they go, because the conductor has resistance to the electrical current flow. If the current in the acupuncture system was as low as that in the output (or growth-stimulating) side of the growth control system, then it could not be expected to go more than a few inches before it would disappear completely.

Engineers handle this problem in telephone cable systems by inserting booster amplifiers along the course of the cable to boost the signal so that it can continue on its way to the next amplifier, and so on. I proposed that the acupuncture points were just such booster amplifiers, spaced along the course of the meridian transmission lines. Further, because we had shown that the DC in the salamander's regenerating limb came from the neuroepidermal junctions, I proposed that the acupuncture points might be the same sort of structure. This seemed to make a great deal of sense in that the skin is where the body meets the environment and where injuries occur. Logically, the detectors for injury *must* be located here.

Since they transmit the signal of injury by DC electrical currents, metallic acupuncture needles inserted in or near such a point would produce sufficient electrical disturbance that the amplifier could not operate, and pain would be blocked. We did some preliminary experiments and found that the real acupuncture points did produce DC potentials. Not only had we shown that the acupuncture points and meridians really existed, but we also had a good theory of what they were and how they worked. We proposed a number of crucial follow-up experiments to NIH, but they did not renew our grant.

THE MISSING LINK: THE INTERNAL DIRECT-CURRENT ELECTRICAL CONTROL SYSTEM

As a result of the DC measurements we had made up to that time, the basic outlines of the complete DC electrical growth control system fell into place. *Input* DC electrical signals carried the information that injury had occurred along the acupuncture meridians to the brain, where part of this group of signals reached consciousness and were perceived as pain. The remainder went to more primitive portions of the brain, where they stimulated similar *output* DC signals that caused the cells and chemical mechanisms at the site of injury to produce repair. This is a complete closed-loop, negative-feedback control system, exactly as I had postulated years before.

We had shown the outlines of an internal, electrical control system that regulated the repair of injuries. While this reintroduced electrical energy into the living organism as a controlling mechanism, this energy was not mysterious and unknowable. Instead, it was an energy that could be measured, and perhaps even manipu-

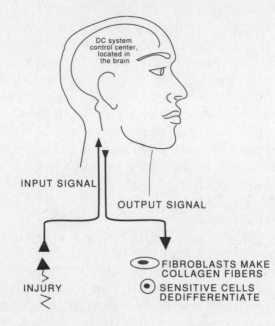

FIGURE 2-7. *The complete closed-loop, negative-feedback growth control system. The input side is the acupuncture system of points and meridians. In the interest of clarity, the "points" that are the booster amplifiers of the acupuncture system are omitted. At this time the nature of the output system was uncertain, except for the fact that it had to be connected with the nervous system as we know it.*

lated, by us. For this reason, a door had been opened to a profoundly important biological control process, one with implications for clinical medicine that extend far beyond the control of normal growth.

CANCER: GROWTH OUT OF CONTROL

Unfortunately, the normal growth of the fetus and the healing of cuts and fractures are not the only kinds of growth that human beings experience. Unlike the salamander, we suffer from the incidence of abnormal growths that can destroy us. Of all diseases, cancer is the most dreaded; it is like an alien invader that inexorably eats away the tissues of the body.

Over the past thirty years, great advances have been made by molecular biologists, and today we know much more about the mechanism of cancer than ever before. Despite this, we have not made similar advances in either the prevention or the treatment of this disease.

The difference between normal cell growth and cancerous growth is simply in the control exerted over the growth process. Embryonic growth and adult healing are precisely regulated by control systems that ensure that growth will be appropriate for the total organism. Cancer is the unregulated growth of cells after they have escaped these controls. What are the controls? How are they exerted? What is wrong with the controls when a cancer grows? Is the problem that of deficient control systems, or is it that the cells have changed so that they no longer respond to normal controls?

The Mechanism of Cancer

Molecular biology has given us the answer to one of these questions: we now know that the defect in cancer is in the cancer cell itself. Those agents that cause cancer (either chemical carcinogens or ionizing radiation) act by changing the genetic apparatus of the cell. Once this apparatus has been changed, the cell no longer responds to the normal control systems of the body.

This discovery was actually made more than eighty years ago. In the late 1800s, Martinus Willem Beijerinck, a Dutch botanist, found that he could transmit a disease of tobacco plants known as "mosaic leaf" by grinding up the leaves, filtering the juice through an ultrafine filter, and then injecting the juice into healthy plants. However, the filter was so fine that it kept back even the tiniest particles, including all bacteria; there was no known living thing so small that it could go through the filter. Beijerinck was convinced that he had discovered a new form of life, which he called a "filterable virus," so small that even the most powerful microscopes could not see it.

No one did anything about this strange discovery until 1909, when a chicken breeder showed up at the Rockefeller Institute for Medical Research in New York with a chicken that appeared to have a cancerous tumor in its breast. The chicken subsequently died, and Dr. Peyton Rous, a pathologist at the institute, performed an autopsy on it. He found that the cancer actually was a sarcoma, a cancer of the connective tissues, and that it had spread throughout the chicken's body, causing its death.

Rous ground up some of the tumor, passed it through the same kind of filter used by Beijerinck, and injected the material into a healthy chicken. This chicken promptly developed the same kind of cancer. Rous was smart enough to realize that if he said, "This cancer is caused by a virus," not only would no one believe him, but his reputation would be forever tarnished. So he instead referred to the "virus" as a "tumor agent." However, Rous's plan did not work; no one believed him anyway, and for more than fifty years the possibility that cancer was caused by viruses was never fully explored. However, the sarcoma cells Rous had isolated grew very well in culture, and they were propagated in this way throughout those years. With the discovery of the electron microscope, viruses were able to be seen, including both the tobacco mosaic virus and the Rous sarcoma virus.

We know now that the virus does its work by incorporating its DNA into that of the cell it infects. The purpose of this is to make the cell produce more virus, but in the process there is a side effect. Some virus DNA contains genes, called oncogenes, that cause cancer in the cells they infect. Research on this finally led to the discovery that normal human DNA contains oncogenes that are normally "repressed," and that chromosomal abnormalities (or mutations) produced when the cells divide can "activate" these oncogenes, subsequently causing cancer. Not all human cancers are caused by viruses, but the evidence is becoming quite clear that *all* cancers are probably the result of this type of mutation in the DNA apparatus.

The mechanism that causes cancer is a piece of DNA. This is, without a doubt, a triumph of technological medicine. But has it brought us any closer to a real cure for this ravaging disease? Unfortunately, at this time the answer is no. The reason for this lies in the mechanistic view that the living cell has a very limited ability to change or heal itself. Molecular biologists seem to view the cell as simply a container for DNA. We have already discussed the ability of cells to defy previous dogma by returning to their embryonic state in regeneration. It now appears that this process has a lot to do with cancer. Unfortunately, this promising avenue of exploration has not been followed. For this story we must return again to the late 1800s.

Regeneration and Cancer

Since cancer is characterized by unlimited growth of cells, one must ask whether this is solely the result of the oncogene activity, or whether the oncogenes produce some more general change in the entire cell. One of the prime characteristics of cancer cells is that they are partially dedifferentiated, with the usual cancer cell being somewhat more like an embryonic cell than an adult one. However, the cancer cell is *not* as dedifferentiated as the cells in the regenerating blastema. It appears possible that this aspect of the cancer cell is what enables it to grow in an unlimited fashion and escape control of the normal body systems. Provided with an appropriate environment and a continuous supply of food, cancer cells are truly immortal. While this is not necessarily a desirable thing, I view it as a miracle: under this circumstance, life can evade death. Certainly, there is something to be learned from this ability, and from the relationship between cancer and differentiation.

In 1877, Julius Fredrich Cohenheim, one of the leading German pathologists of the time, suggested some interesting ideas on cancer. He was the first to observe that the cells of cancers were less mature than the cells from which they were derived. To him they appeared to be almost, but not quite, like the cells in the embryo. Cohenheim proposed that cancers were caused by groups of embryonic cells, which he called "embryonic rests," that were left over in the adult body. To find out why these resulted in cancerous growth, he made a startling conceptual leap. He proposed that there were control systems in living things that regulated cell growth, and that the control system in the adult was different from that in the embryo. The result was that the embryonic cells left over in the adult would not be regulated by the adult control system and would then grow in the uncontrolled fashion characteristic of cancer. More importantly, Cohenheim postulated that if the embryonic control system could be produced in a cancer, the cancerous cells would revert back to the totally embryonic state and then return, by the normal process of differentiation, as *normal* cells. Cohenheim's "embryonic rest" theory is, of course, wrong, and his idea that the cancer cell could revert back to a normal cell is very much at odds with present knowledge.

Molecular biologists look at the DNA code in the cell as something fixed and immutable, except when an outside agent such as a virus or carcinogen interacts with the DNA itself. Therefore, anything that would have its effect upon the cell itself could not alter the DNA. To these scientists, then, the idea of cancer cells becoming normal is ridiculous; the only way to control cancer is to kill it or to find a way to actually enter its DNA and alter it.

Fortunately, long before we knew all this, a few brave souls looked at Cohenheim's "crazy" ideas. In the 1920s and 1930s, a few experiments showed that chick embryo cells implanted into adult chickens became cancerous. However, these tests were so far outside the mainstream of scientific thought that no one paid much attention to them. In 1948 a biologist friend of mine, Dr. S. Meryl Rose, conceived of a way to test Cohenheim's thesis. Rose postulated that when the salamander regenerated a limb, the original embryonic control systems were locally reactivated, producing dedifferentiation of the mature cells in the wound. The resultant regenerative growth was then identical to a "replay" of the original embryonic growth of the same limb. The possibility occurred to Rose that if cancer cells were implanted into a regenerating salamander's limb, they could become normal—if Cohenheim was right.

Since salamanders never normally develop cancer, Rose's first experiment was to transplant frog kidney-tumor cells (now known to be caused by a virus) into the foreleg of some salamanders to make sure that they would "take" and act like a typical cancer. He found that the cells did become incorporated into the salamander, growing locally for a while as a visible tumor and then spreading in typical cancer fashion to the remainder of the body, ultimately killing the salamander.

After Rose had proved that frog kidney-tumor cells were cancerous to salamanders, he repeated the experiment. But this time, when the cells grew into the local tumor, he amputated the salamander's foreleg *through* the tumor, leaving cancer cells behind in the amputation stump. As the salamander began to regenerate the missing portion of its foreleg, the cancerous tumor mass disappeared, merging with the developing blastema. When the forelegs were regenerated, they were normal—and the tumor was gone. There was no evidence of cancer, and the animal lived normally thereafter.

When the regenerated portions of the forelegs were examined microscopically, Rose could see nuclei of the original frog cancer cells in the normal salamander cells of bone, muscle, and so on in the regenerate. He concluded that in the presence of a regenerative growth process, cancer cells could revert back to normal cells and become part of the growing blastema. In other words, Cohenheim had been right. *Cancer cells were not locked irretrievably into the malignant state; in the presence of embryonic control systems, they could return to normal.* Unfortunately, this work was still outside of the mainstream. Only a few scientists have duplicated this experiment since then. While they have all reported the same results that Rose did, no one has paid any attention.

Clearly, *something* in the process of regeneration has the ability to speak to the cancer cell and cause it to rearrange its genetic apparatus so as to deactivate the oncogenes. Possibly, the signal to dedifferentiate is stronger than the oncogene signal to reproduce. The fact that DC electrical currents play a significant role in regeneration—and also that cancers have extremely high DC electrical potentials—would seem to indicate that electrical factors may be involved. Some attempts have been made to study this idea, and in later chapters we will discuss the relationship between electricity and cancer.

Since cancer cells can become normal cells in this fashion, shouldn't we be studying this process as a biological way to control this disease, rather than continuing to try to kill the cancer cells? In a guest editorial in the journal *Growth*, Dr. Alexander Wolsky strongly suggested a return to basic biological research on the relationship between cancer and regeneration. He also spoke of the "Vietnamization" of the war on cancer—the expenditure of enormous amounts of money with no effect on the enormous losses of life. Clearly, it is time for a change in outlook.

Cancer and the Rest of the Body

The discovery of the mechanism behind cancer seems to have done little to dispel the idea that cancer is somehow something foreign within the body and that the body has little to do with it. This notion, coupled with Ehrlich's idea of a "magic bullet" for each disease, has led medical science down the road of devising chemical agents to kill cancer cells. This is not an easy task; the cancer cells, changed though they are, are still *our* cells. Not only are chemotherapeutic agents against cancer toxic to the rest of the body, producing depression of the immune system, but many of them are themselves carcinogens. Only in recent years has the idea that the body has defenses of its own against cancer become popular. The roots of this concept are also buried in the late 1800s.

At that time, Dr. William Coley, a surgeon at New York's Memorial Hospital, noted another puzzling fact about cancer. Many of his patients with cancer became completely cured if they developed a very serious life-threatening infection and survived. Coley postulated that the infection had activated immune resistance mechanisms in the patient and that, after the infection had literally brought the patient to death's door, the activated immune system not only successfully fought off the infection but destroyed the cancer as well.

Coley mixed up a combination of the worst bacterial toxins he could find and began injecting them into patients with inoperable cancers. Some died from the "treatment" and some did not show any remission of the cancer, but many went on to a complete cure. In 1975, a review of 186 patients treated with Coley's toxin showed cures in 105 cases, a much better success rate than any other treatment then, or now, available.

Despite these promising results, no large-scale clinical studies have ever been done, and Coley's idea has been followed up in only a few laboratories. This laboratory work, however, has centered on the mechanistic concept that either the bacteria or the body must have produced one specific chemical that killed the cancer cells selectively, without harming the body's normal cells. This work has led to the discovery of a chemical in the bacteria's cell membrane that causes the body to produce another chemical, one that when injected into mice with implanted cancer results in the death of the tumor cells alone. The tumors turn black and liquefy. The chemical involved is called tumor necrosis factor (TNF). At present, there is considerable interest in producing this chemical by means of recombinant-DNA techniques so that it can be used in human patients with cancer.

However, there are problems with this approach. First of all, Coley's patients experienced a *disappearance* of their tumors; they did not turn black and fall off. TNF thus appears to be something else entirely. It has been noted recently that TNF has been studied in other laboratories where it has been given another name—endotoxin or cachexin. In this work, endotoxin was associated with the extremely serious and often fatal clinical condition known as toxic shock syndrome. In this condition, profound shock and death occur within a few hours of the onset of a bacterial infection. The chemical responsible is endotoxin (cachexin)—the same chemical as TNF.

So the chemical pathway that is now being followed has produced another single agent that appears to be even more toxic to the human body than the synthetic anticancer drugs now in use. Why this chemical produces acute toxic shock syndrome in one case and necrosis of tumor cells in another is indicative only of the fact that other factors within the body are operating.

Advocates of scientific medicine still seem to believe that "once a cancer cell, always a cancer cell," so the only treatment alternatives possible are to kill the cell (preferably with a single chemical agent) or to surgically remove it. The vital role played by the body's own resistance factors is still studiously ignored in this approach. It seems to me to be the height of folly to treat cancer with agents that decrease these innate resistance factors.

Resistance to cancer by the immune system is not the only mystery left unexplored by modern technological research. Later on, we will discuss such strange occurrences as "spontaneous remissions," in which, for no apparent reason, cancers disappear. And we will explore the profoundly important role played by psychological factors in cancer.

At the very least, the ability of cancer cells to revert to normal and the innate resistance factors of the total body should be thoroughly investigated as possible biological cures for this highly feared disease. Such an approach would seem imperative in view of a recent thought-provoking article, "Progress against Cancer?" by Dr. John C. Bailer, a respected medical statistician, and E. M. Smith. "We are losing the war against cancer," Bailer writes, basing this conclusion on a thorough statistical analysis of the years between 1950 and 1982. That analysis showed the incidence of cancer to have increased steadily, despite improvements in diagnosis and treatment.

The mysteries of growth seem to be yielding to the concepts of electronic biological control systems. We are now beginning to understand how almost vanishingly small electrical currents act upon certain cells to produce healing and, in some instances, to cause cancer cells to return to normal. While this is of obvious importance in clinical medicine, providing us with the ability to actually manipulate the growth process at will, it is of even greater importance to the success of the control system theory, which can be applied to many other areas of life. This is not a return to mysterious vitalism, but rather a logical extension of the mechanistic concept that reintroduces the ancient idea of life energy in a scientifically valid fashion.

THE MORPHOGENETIC FIELD: BLUEPRINT FOR GROWTH

After both the input signals of injury and the output signals that stimulate healing were found to be parts of an electrical control system, we were left with a problem: How was the new growth "told" what kind of tissue to become? A simple cut on the finger is easy; all that's needed is to close it by fibrosis. But consider the salamander's regeneration of a limb. Not only is it necessary for something to tell the blastema what to become, but the veritable mountain of information characterizing that specific extremity has to be transmitted to it as well. As I noted before, the salamander's foreleg is the counterpart of the human arm, right down to the tiny bones in its "fingers." All the information on what cells go into the organized tissues, how the tissues fit together, on the bones, joints, nerves, ligaments, blood vessels, and so on has to be transmitted to the site of injury. Since the instructions relate the new growth to the total body, the system providing the instructions has to be *part* of the total body. What is it, and where is it? Is it part of the brain?

The nineteenth-century embryologists recognized this problem. In an attempt to solve it, they devised a few simple experiments concerning the blastema. What they found revealed an amazing and even more inexplicable complexity.

It is a simple matter to remove the blastema and transplant it somewhere else on the salamander's body, where it will then grow. If the scientists removed the blastema from a foreleg amputation before the tenth day after amputation and transplanted it near the intact hind leg, it would grow into a hind leg. The same blastema transplanted next to the tail became a tail. The embryologists thus concluded that there were "body-inductive areas," or "morphogenetic fields," that contained the information that specified what was to grow in that area. These experiments revealed a very important fact: There is some mechanism within the living salamander that contains an overall plan for the salamander's body and provides the information that instructs the blastema what tissue it should construct.

That was a monumentally important piece of information, but other experiments made it even more important. The embryologists next found that if they waited until *after* the tenth day to do a transplant, the foreleg blastema became a foreleg, *no matter where it was transplanted.* Somehow, in a very short period of time, the necessary instructions to make a terribly complex structure had been transmitted to this unimpressive little lump of primitive cells. To make matters even more complicated, it was found that the nervous system—the only means then known for transmitting such information—was not involved in the normal fashion. While the local nerves would grow into the transplant before it began to grow itself, both these nerves and those going to the amputation stump were completely "silent" throughout this period of information transfer—that is, they transmitted no information by means of nerve impulses.

It was obvious that chemical messengers were totally unable to convey this much organizational complexity. Since the nervous system was also excluded, there had to be something else—something similar to the morphogenetic field—that could contain within itself the entire organizational plan. Since DC electrical currents had been shown to be deeply involved in regeneration, could such a field produced by such a current contain this information? The answer is a qualified yes. Even the simplest electrical field has a high degree of organizational complexity.

If, for example, you place two electrodes in a petri dish filled with a mild salt solution and then connect them to a small battery, a current will flow between the electrodes. When you look at the pattern of the current itself, it is not simply a stream of charge carriers running directly between the two electrodes, but a spread-out *field* composed of changing values of current and voltage.

A large amount of information may be contained in a field such as this. Each point in the field may be characterized by a unique set of values of voltage, current strength, and current direction. While each point would therefore have a unique set of coordinates, this in itself is not enough to contain the entire "blueprint" for a complex extremity. The field provides only positional information. However, the field must act upon the primitive cells of the blastema, each of which contains a large amount of information. It's possible that the field serves only as a "trigger" to instruct the cells to follow certain routines in development. However, the electrical-field concept of the morphogenetic field would seem to be at least a good starting point from which to embark on a course of study.

If local electrical currents and voltages controlling healing are

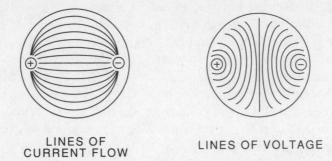

LINES OF   LINES OF VOLTAGE
CURRENT FLOW

FIGURE 2-8. *In the diagram on the left, the current is shown to flow as*
a series of streamlines from the positive electrode to the negative. In
the diagram on the right, the lines of equal voltage, also called
equipotential lines, are shown. On each voltage line the value of the
voltage is the same, but the value for each line is different from the
others. Therefore, for any single point in the voltage/current field
there will be a unique set of values of voltage, current strength, and
current direction.

not simply local phenomena but are instead part of a larger, total-
body DC system serving as the morphogenetic field, there must be
a *second nervous system,* one that controls the most primitive functions and that probably predates the nervous system with which we
are so familiar. The existence of such a system would force us to
change our ideas of how the brain works.

THE DUAL NERVOUS SYSTEM UNCOVERED: ANALOG AND DIGITAL CONTROLS

It sounds quite revolutionary—and perhaps even rather
"fringey"—to say that there is a second, hidden part to the brain.
While we do know a great deal about how the brain and nervous
system work, there are some major questions that have not yet been
answered. I have always believed that our most important roadblock
in this area is that we consider all brain functions to be based on one
mechanism, the nerve impulse. This is a very sophisticated mechanism in its own right; when it is used to transmit information, it
becomes an extremely complicated mechanism.

The best way to approach this problem is to compare the currently accepted mechanism of information transfer in the nervous

system to a digital computer. In both the computer and the brain, the basic signal is a digital one, a single pulse. The information is coded by the number of pulses per unit of time, where the pulses go, and whether or not there are more than one channel of pulses feeding into an area. Our senses—smell, taste, hearing, sight, and touch—work by means of this type of pulse. The digital computer on my desk works in the same way. This digital system works extremely fast, and it can transmit large amounts of information as digital "bits" of data.

I have difficulty believing that the first living organisms used this same type of system. It would seem that they must have had a much simpler mechanism for transmitting information. The first living organism did not have eyes, ears, or other such sensory organs, and it wasn't much concerned with thinking great thoughts. What it needed was a way to know when it was injured and the means to efficiently repair itself, along with the ability to relate to the important aspects of its environment and other living organisms. In short, it must have been rather unimpressive and totally lacking in complexity. Since there was no requirement for quick action or for the transmission of large amounts of sophisticated information, it could make do with a much simpler system of *analog* data transmission and control.

Before digital computers, there were analog computers. These operated in exactly the way I believe the first living organisms did. The analog computer works by means of simple DC currents, with the information coded by the strength of the current, its direction of flow, and slow, wavelike variations in its strength. While this system is slow and is incapable of transmitting large amounts of data, it is extremely precise and works well for what it is designed to do. Today, some of the most sophisticated computers are actually hybrid computers, containing both analog and digital components.

If the first living organisms used an analog type of data-transmission system and if such a system is still used for injury repair, then the entire analog system must be present—although unseen— as part of our entire data-transmission and control system. The brain may thus operate as a hybrid computer, with a very important analog part that operates by means of DC electrical currents. This concept provides us with an entirely new way to look at the old brain/mind/body conundrum.

My first experiments involved searching for DC electrical potentials in the uninjured salamander. Similar searches had been conducted in the past, notably by Professor Harold S. Burr at Yale in the 1920s and 1930s. Using the best measuring devices available, he found DC potentials organized in a simple "front-to-back" fashion, with the salamander's nose being negative and the tip of the tail positive. When I began my work some years later, I benefited from the technical advances in instrumentation that had been made during World War II. Consequently, I found a more complex pattern that appeared to be related to the total nervous system, from the brain to the end of the furthest nerve fiber.

The nerve cells of the salamander are concentrated in three places: the brain, the cervical enlargement, and the lumbar enlargement. All are connected by the spinal cord, and from each enlargement nerve fibers go out to the periphery. Some of the fibers from the brain go to the special sense organs, but most go down the spinal cord to the other two enlargements. Fibers from the cervical enlargement go to the forelegs, while those from the lumbar enlargement go to the hind legs, with a few to the tail. The arrangement of our own central nervous system is basically the same.

We found that in the awake, unanesthetized salamander, the three collections of nerve cells were electrically positive, and the nerves coming from them were more and more negative as our measuring electrode passed further out along the limbs. The DC potentials seemed to be produced by the nerve cells themselves, because cutting the nerves to a limb in the lightly anesthetized salamander reduced the DC potentials in that limb to zero. However, the level of activity of the nerve cells seemed to be related to the state of the entire DC system. For example, general anesthesia resulted in drops in the DC levels, starting with the brain and then proceeding throughout the rest of the nervous system. Finally, in deep anesthesia, all potentials were zero or were even slightly reversed.

We then found that the direction of flow of the DC current in the motor nerves (those that make muscles move) was most often opposite to that in the sensory nerves. However, in both cases we were able to show that the electrical current was semiconducting and not ionic.

The method we used demonstrates a unique relationship between semiconducting electrical currents and magnetic fields, one that will later be shown to be important. If there is an electrical current flowing between point A and point B and you put a strong magnetic field at right angles to the line of current flow, there will

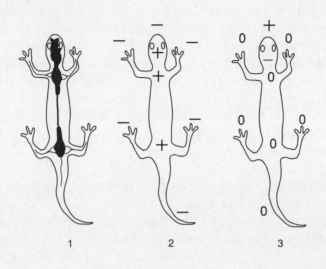

FIGURE 2-9. *Number 1 shows the arrangement of the major parts of the salamander's central nervous system—the brain and the enlargements of the spinal cord, where the nerve cells send fibers to the limbs. Number 2 shows the pattern of DC electrical potentials we found. These seemed related to the nervous system because each concentration of nerve cells was positive, while the nerve endings were negative. Number 3 shows the changes in this pattern of electrical potentials that occurred with general chemical anesthesia.*

be a "pull" on the charge carriers from the magnetic field. If the current is electrons flowing along a wire, nothing much will happen because the electrons are tightly held, or constrained, within the metal. If the current is ionic, again, nothing much will happen (unless you use an enormously strong magnetic field), because the ions are large objects and move slowly. However, if the current flow is composed of electrons or holes in a semiconductor, the charge carriers move very easily, and many will "pile up" in the magnetic field. If you then put two electrodes on the conductor so that the piled-up electrons can flow into another circuit, a respectable voltage called the "Hall voltage" can be measured in that circuit. I measured respectable Hall voltages from a salamander's forelegs, voltages that could have come only from a semiconducting current carried by the nerves.

So there really is a semiconducting DC system operating, at least in the nerve fibers in the limbs. But what of the brain?

Looking back in the scientific literature, I found that in the 1930s several eminent neurophysiologists had measured DC voltages in the brain and had shown that these actually controlled the level of activity of the nerve cells in the brain. While this work was of obvious importance to the understanding of how the brain worked, it was essentially lost as instruments for easily and accurately measuring the nerve impulses came into use and the difficult techniques needed for measuring DC voltages fell into disuse. Besides, everyone "knew" that the nerve impulse was where the action was, so why waste time measuring something that didn't make any sense? I decided to reopen this controversy, but from the point of view of the total-body DC field.

Since general anesthesia produced a drop in the normal front-to-back negative-DC field across the brain first and then through the rest of the nervous system, I postulated that the chemical anesthetic actually worked by causing this change in the DC system. Therefore, it should be possible to anesthetize a salamander simply by placing electrodes on the front and back of its head and passing a small current through the head in a back-to-front direction. In order to prove that the animal was anesthetized and not "electrocuted" (even though I was going to use a very small current), I decided to measure the salamander's brain waves before and during the time the current was on. It was very exciting to watch the electroencephalogram (EEG) change from the typical "awake" pattern of eight to ten pulses per second to the slow waves of two to three per second that are characteristic of deep anesthesia, when only a few *micro*amperes of current were applied with the frontal electrode positive.

Later, I was able to show that the currents in the brain were, like the currents in the peripheral nerves, semiconducting. With no external source of reverse current applied to the head, a magnetic field of 3000 gauss, applied at right angles to the front-to-back line of normal current flow, completely stopped the electrical current in the brain and produced the identical slow-wave pattern in the EEG.

A DC analog system of data transmission and control is, in fact, hidden within the central nervous system. It had seemed unlikely to neurophysiologists in the 1930s that nerve cells had within themselves two such completely different electrical mechanisms of action. They therefore postulated that it was the other cells of the brain, the perineural cells, that generated this DC electrical activity.

We commonly think of the brain as being packed with billions of nerve cells. While there *are* billions of nerve cells in the brain, there are as many, if not more, perineural cells. In fact, these cells are found wherever nerve cells or nerve fibers are found. They come in a variety of types, the most common of which are the glial cells in the brain and spinal cord and the Schwann cells encasing the peripheral nerves. In the brain, the nerve cells can be likened to "raisins" in a "pudding" of perineural cells.

All the perineural cells are related and arise, embryologically, from the same basic cell line (the ectoderm) as the nerve cells and skin cells. Anatomically, the perineural cells appear to be a system of their own.

The Schwann cell surrounding a fine, terminal branch of the nerve in the fingertip is in contact, via the other Schwann cells along the entire nerve, with the Schwann cells in the spinal cord, which in turn are in contact with the glial cells of the brain. Every nerve cell or nerve fiber is surrounded by perineural cells of one type or another. If there were a way to dissolve the nerve cells, you would still see the brain, spinal cord, and nerves apparently intact.

The function of the perineural cells has been something of a mystery. Most theories have ranged from seeing these cells as the packing material that protects nerve cells, to seeing them as providers of nourishment to the nerves. However, it has recently been shown that the perineural cells can generate electrical potentials and pass them along from one to another. The possibility has never been seriously considered that the perineural-cell system, extending throughout the whole body exactly as the nerves do, functions as a primitive communication system.

UNCOVERING THE HIDDEN CELLULAR-COMMUNICATION SYSTEM

The relationship between the perineural cells and the DC system was discovered accidentally, after one of my students became interested in the nonunions of fractures. Strangely, this seemed to be a condition unique to human beings, because it was practically impossible to produce the same thing in animals. This student thought that if he fractured the tibia of a rat and at the same time cut the nerve supply to the leg, a nonunion of the fracture might result.

The first experiment was a failure. The fractures in the rats healed only a few days slower than normal. I suggested to the student that in the next experiment he cut the nerve supply two days before fracturing the tibia. This group of rats healed their fractures slightly faster than the first group, but still a little slower than normal. This was getting interesting, so we did a third experiment in which the nerve supply was cut five days before the fracture. These rats healed their fractures in the normal time!

Obviously, something had happened during the five days that had restored normal healing. It could not have been a regrowth of the nerve to the leg, because this would take at least one month, and the rats in the third experiment had no functional use of these limbs even after the fractures had healed. Fortunately, the student was a good investigator, and he had saved the entire legs of the animals as experimental specimens. We went back and dissected out the nerve supply in the third experimental group. We found that while the nerve had just begun to grow, there was a strand of *something* connecting across the gap in the nerve. Under the microscope, we found that this strand was composed of tubes of Schwann cells that would later contain the regrown individual nerve fibers.

The perineural cells—in this case, the Schwann cells—carry the electrical signals that cause fractures to heal. The nerve has nothing to do with it. It seemed justified to postulate that the DC system we were studying resided in the perineural system. If this were so, then the perineural cells were more primitive than the nerve cells and likely represented the more primitive analog data-transmission and control system. However, this implied that the perineural cells were the precursors for the nerve cells, and this was totally against the mechanistic idea that cell lines were fixed and non-changeable.

I decided to do a simple experiment. I grew perineural cells in culture, alone, with no nerve cells present. After a week or so, these cells had grown into interesting swirling patterns. When I looked carefully, I could see an occasional typical *nerve* cell mixed in with the perineural cells. While this does not absolutely prove that the perineural cells can transform into nerve cells, it does tend to support this notion.

Interest in the DC system, particularly in its interrelationships with the nerve cells in the brain, has been increasing over the past few years. Recent experiments have shown that when a human subject is told to make a certain muscular movement after being given a signal, there is an increase in negative DC after the signal, but that this occurs almost a half-second *before* the muscular action is performed. It appears that the DC is somehow involved in getting the neurons ready to fire the command to move the muscles. This phenomenon, which has become known as the "readiness potential," seems to imply that the DC system *commands* the nerve-impulse system.

In a recent review article, Dr. Benjamin Libet of the University of California, a long-time researcher in the DC-potential area, presented even more startling data. He was concerned with which came first: the readiness-potential shift, or the decision itself. Libet's experiments indicated that the readiness potential *preceded* the decision. In his words, "The brain seemed to have a 'mind' of its own." However one wishes to interpret this, it seems clear that the DC-potential system in the brain is activated prior to the nerve-impulse system, and that the latter may depend upon the former's being in a particular electronic state. The DC system thus appears to be, in fact, the place where the actual command decision is made.

If we combine all of the above observations, we can only conclude that a more primitive, analog data-transmission and control system still exists in the body, located in the perineural cells and transmitting information by means of the flow of a semiconducting DC electrical current. This system appears to have been the original data-transmission and control system present in the earliest living organisms. It senses injury and controls repair, and it may serve as the morphogenetic field itself. It controls the activity of body cells by producing specific DC electrical environments in their vicinity. It also appears to be the primary system in the brain, controlling the actions of the neurons in their generation and receipt of nerve impulses. In this fashion it regulates our level of consciousness and appears to be related to decision-making processes.

This startling conclusion leads to more questions:

Is the analog DC system "smarter" than the nerve-impulse system? Does it regulate and control the digital nerve-impulse system in the same fashion that we control and use our personal computers? Is it the seat of thought, logic, and memory? Can we control its operations and heal ourselves?

These are intriguing questions, but before we can answer any of them, we have to add some unexpected and important complexities to this dual brain system: the magnetic connection.

CHAPTER THREE

THE NEW SCIENTIFIC REVOLUTION: THE MAGNETIC CONNECTION

Long distance cell-to-cell or organism-to-organism communications may be accomplished by transmission and reception of electromagnetic signals through membrane receptors or enzymes.

TIAN Y. TSONG, M.D., "Deciphering the Language of Cells," in *Trends in Biological Science*

*O*ver the past decade, important discoveries have been made indicating that living things sense and derive information from the natural magnetic field of the Earth. While I had postulated this in the early 1960s and had theorized that the mechanism involved was the interaction of the geomagnetic field with the system of internal DC currents, I was only half right. The mechanisms of interaction are much more complex and sophisticated and involve actual anatomical structures "designed" for this purpose.

MAGNETIC FIELDS AND BIOLOGY

When I made my discoveries in the early 1960s, physics had shown that any flowing electrical current produced a magnetic field in the space around it, and that a changing, pulsing, or moving magnetic field could produce an electrical current flow in an electrical conductor placed within the field.

In its simplest form, a magnetic field is a field of magnetic force extending out from a permanent magnet. Everyone is familiar with the illustrations of the lines of magnetic force around a bar magnet. Magnetic fields are produced by moving electrical currents, and the field of the bar magnet is produced by the spinning of the electrons around the atomic nuclei in the iron material of the bar magnet itself. In order to act as a "magnet," these atoms must be lined up in the same direction so that the individual magnetic fields combine to produce one big field. If a magnet is dropped on a hard floor, the magnetic field is destroyed because the atoms are "shaken up" and their direction is changed, resulting in a random pattern.

When an electrical current flows in a wire, the movement of the electrons produces a similar field in space, but one that is oriented *around* the wire. If the current is a DC current, the magnetic field is steady, like that from a permanent magnet. The strength of the magnetic field depends upon the amount of current flowing in the wire—the more current there is, the stronger the magnetic field will be. Very large DC current "electromagnets" are used to pick up heavy metal objects.

All magnetic fields have a direction, or vector. We know that the bar magnet has a "north" pole and a "south" pole. These names, coined long ago, reflect the fact that unlike poles attract each other, while like poles repel each other. The pole of the bar magnet that pointed to the Earth's North Pole was called the "north" pole. Obviously, the terms "north" and "south" have no special meaning except to indicate that the magnetic field has direction as well as strength.

If the electrical current in the wire is fluctuating, the magnetic field will have the same fluctuations. We characterize the field in this case on the basis of the rate, or frequency, of fluctuation (for example, once per second, a thousand times per second, etc.). In science we use the term *hertz* (Hz) rather than *per second*, after Heinrich Hertz, who first studied this phenomenon. Once per second thus becomes a field of 1-Hz frequency, and so forth.

Now we come to a difficulty. The field fluctuating in this fashion theoretically extends out in space to the end of the universe, but decreasing in strength with distance and ultimately becoming lost in the welter of other magnetic fields that fill space. It is called an electromagnetic field because it contains *both* an electrical field and a magnetic field. Since it is fluctuating at a certain frequency, it also has a wave motion. The speed with which it moves outward is the same as that of light (roughly 186,000 miles per second). As a result, it has a wavelength, which depends upon the frequency. For example, a 1-Hz frequency has a wavelength in millions of miles; a 1 million-Hz frequency (shortened to one megahertz, or one MHz) has a wavelength of several hundred feet; and a 100 million-Hz frequency (100 MHz) has a wavelength of about six feet.

All of the possible frequencies of electromagnetic waves or fields can be put onto an electromagnetic spectrum, starting with the slowest frequencies and going to the highest. The electromagnetic spectrum is usually pictured as a line that increases from zero on the left to trillions of cycles per second on the right. About three-fourths up the frequency spectrum is something we are very familiar with—light. Sunlight, as well as the light from electric lights, is the same thing as the magnetic field. It is produced by the movement of electrons and has the same characteristics as the Earth's magnetic field, radio waves, and X-rays. It is an electromagnetic field, but one for which we have developed specific anatomical detectors—our eyes.

All electromagnetic fields are force fields, carrying energy and capable of producing an action at a distance. It would appear that we know everything about electromagnetic fields. But we don't. One of the big problems of physics is the fact that these fields have characteristics of both waves and particles. Depending on how you look at light, radio waves, or any other part of the electromagnetic spectrum, you will find either waves or particles, called photons. The photon is a very strange thing: it is a particle, a little piece of something, but it has no mass; no one has ever seen one. In quantum theory, photons carry the energy. Those with higher frequencies have more energy than those with lower frequencies. In the human eye, one photon gives up its energy to the retina, which somehow converts it into the electrical signal that produces the sensation of light.

In sum, magnetic and electromagnetic fields have energy, can carry information, and are produced by electrical currents. When we talk about electrical currents flowing in living organisms, we also imply that they are producing magnetic fields that extend outside of the body and can be influenced by external magnetic fields as well.

In the early 1960s, I predicted that the flow of this DC current in the brain would produce a magnetic field that could be observed at a distance *outside* of the head if one had a magnetometer of sufficient sensitivity. When I made this statement at a scientific meeting, there was considerable laughter from the audience. I was told that such a device would never be made, and that even if such a magnetic field could be measured, its strength would be so small that it would be of no physiological consequence whatsoever.

It was not until several discoveries were made in solid-state physics and electronics, culminating in the design of the SQUID (superconducting quantum interference detector) magnetometer in 1970, that I was proved correct. Today, the magnetic field (called the magnetoencephalogram, or MEG) produced by the brain is easily detected using the SQUID magnetometer (a device based upon the fact that a superconducting current is extremely sensitive to very small magnetic fields). The flow of the electrical currents in the brain actually does produce a magnetic field that can be measured and analyzed several feet away from the head. It was gratifying to read that a study with the MEG had shown a vector of DC running between the front and back of the brain, exactly as I had discovered many years before by measuring DC potentials alone.

It may be a little disconcerting to know that we, and all other living things, are surrounded by a magnetic field extending out into space from our bodies, and that the fields from the brain reflect what is happening in the brain. The implications of this are enormous, and I will discuss some of them later on.

In the 1960s, I predicted that living organisms could be influenced by external electromagnetic fields via the physical interaction of the fields and the DC electrical currents flowing within the organisms. While this was theoretically true, the very small strength of the currents would require a magnetic-field strength considerably greater than that of the Earth in order to have any material effect upon the organisms. Therefore, the idea that living things could be influenced by changes in the Earth's low-strength geomagnetic field remained an "old wives' tale." However, some biologists obtained excellent evidence of a direct, important relationship between the Earth's magnetic field and the cyclic behavior of living organisms.

LINKING BIOLOGICAL CYCLES TO THE EARTH'S MAGNETIC FIELD

By the 1960s, the phenomenon of biological cycles had been known for some time and had attracted increasing interest and experimentation because it appeared to have some mysterious relationship to certain clinical conditions. For example, many psychological and psychiatric disturbances were characterized by disturbances in the sleep-wakefulness cycle. It was known that living things, including humans, had an intrinsic cycle in sleep-wakefulness activity that occurred even when environmental cues, such as daylight and temperature, were excluded. The mechanists explained this by simply postulating that a chemical oscillator or clock timer for the cycles existed somewhere in the brain.

This idea neglected one important fact: all of the cycles had rates of change that were practically identical to similar cycles in our geophysical environment. The biological cycles were similar not only to the light cycles of day and night, but also to the daily tidal changes in the Earth's magnetic field and to the lunar cycle of twenty-eight days (which was well represented in the pattern of female menstruation). Long-range patterns of change in human affairs had been shown to have some relationship to the twenty-two-year cycle of solar magnetic-field activity, and some disturbances in human behavior had appeared to be related to magnetic storms in the Earth's geomagnetic field. To the mechanists, these links were simply happenstance, because there was no physical mechanism by which any physical field other than light could influence the actions of living organisms.

Nevertheless, this tantalizing relationship led a number of scientists to think the unthinkable: perhaps living organisms detected and derived timing information from the cyclic pattern in the Earth's magnetic field.

Among the scientists intrigued with this possibility was Professor Frank Brown of the Woods Hole Marine Biological Laboratory in Woods Hole, Massachusetts. Brown completed research that established beyond any doubt that the cycles of a number of simpler organisms could be changed in dramatic fashion simply by exposing them to a magnetic field (from a small permanent magnet) of the same strength as the Earth's, but pointing in a different direction.

The strength of the Earth's magnetic field averages about half a gauss, and, worse yet, its daily change in strength is less than 0.1 gauss. Compared with the 200-gauss strength of the permanent magnet that holds a refrigerator door closed, this is peanuts indeed; there is simply no way that such a small change in the field could influence even a compass needle. However, Brown's experiments indicated clearly that living organisms had the ability to somehow sense these minute daily cycles in the Earth's magnetic field and to use them to time their biological cycles. The notion that living things could be influenced by such a weak field was akin to Mesmer's claims, and despite the validity of Brown's experiments and the importance of his data, his findings were simply ignored.

MAGNETIC SENSE ORGANS IN LIVING THINGS

Today, twenty years after Professor Brown's experiments, we know that living things have the capability of detecting and obtaining information from such low-strength fields as the steady geomagnetic field and its cyclic fluctuations. They accomplish this by means of two specific anatomical structures that are connected with the brain. Brown was right.

The "Magnetic Organ"

In 1975, Professor Richard Blakemore, also of the Woods Hole Marine Biological Laboratory, became intrigued by the strange behavior of some bacteria he was studying. Blakemore noticed that the bacteria always clustered at the north side of their culture dish. Even if he turned the dish so that they were at the south end and left it overnight, the next morning the bacteria were back at the north side. While such "magnetotrophic" bacteria had been described before, no one had ever done what Blakemore did next: he looked at them under the electron microscope. What he found was astonishing. Each bacterium contained a chain of tiny magnets! The magnets were actually crystals of the natural magnetic mineral magnetite, the original lodestone of preliterate peoples. Somehow, the bacteria absorbed the soluble components from the water and put them together in their bodies as the insoluble crystalline chain.

Later studies showed that this arrangement was of value to these bacteria, which lived in the mud on the bottom of shallow bays and marshes. If they were moved by the tide or by storm waves, their magnetic chains were large enough (in comparison to their body size) to physically turn their bodies so that they pointed down at an angle corresponding to the direction of magnetic north. All the bacteria had to do was swim in that direction, and sooner or later they would be back in the mud. This was an interesting mechanism, but it did not contain any sophisticated information transfer. The bacteria did not "know" that north was the way to swim; they just did so. However, these observations opened up a much more interesting series of investigations.

Some higher organisms demonstrate amazing abilities of migration and direction-finding. Of these, the homing pigeon has been the most studied. While various attempts had been made to show that these pigeons somehow used a magnetic compass, all had come to naught. Then, in 1971, Dr. William Keeton of Cornell University reported on a lengthy series of experiments.

Keeton conceived the idea that, like commercial airliners, pigeons had more than one navigational system. He proposed that the one used most often was polarized light from the sun, which would give the birds a sort of steady compass direction. This would be used in conjunction with a "map sense," a remembered visual representation of the ground with recognizable landmarks (pilots would call this "visual flight rules," or VFR). Keeton postulated that the birds also had a third, stand-by system, one that would be used only in the absence of the first two (such as in a dense fog). This was the magnetic sense. In a stroke of genius, Keeton found a way to cancel out the first two systems without harming the pigeons or interfering with the experiment: he fitted them with translucent contact lenses. The lenses permitted light to enter, but they prevented the entry of *polarized* light or of any visual images. If the pigeons could still find their way home, Keeton reasoned that it must be with the help of a magnetic sensing system.

He released his pigeons in the Adirondack Mountains of New York State, about 100 miles from Cornell in a straight-line distance. The pigeons with the contact lenses came home just as well as the pigeons without lenses, except that they made a detour. Instead of flying in a straight line to Cornell (as the control pigeons did), they first flew west, *out over Lake Ontario*. Pigeons *never* fly over large bodies of open water, but these pigeons could not, of course, see the lake. At a certain point they turned south and went straight to Cornell, arriving somewhat later than the others. Keeton wondered if they had perhaps flown west until they crossed a line of the Earth's magnetic field that intersected with Cornell.

Keeton's next experiment was the clincher. To the back of the head of each pigeon with contact lenses, he glued a small magnet of about one-half gauss strength (roughly equal to the local geomagnetic field, but oriented opposite to it in direction). When released, these birds scattered, flying in all directions and not returning to Cornell until the contacts fell off, about a day or two later. These birds had a magnetic sense that was canceled out by a small, abnormal magnetic field applied to the brain. So there was no question: pigeons had several navigational systems, and one of them was *an ability to sense, and derive directional information from, the Earth's magnetic field.*

After a humble bacterium had been shown to have a magnet

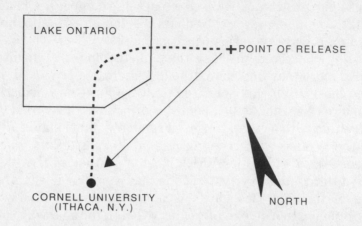

FIGURE 3-1. *The flight path of homing pigeons in Keeton's first experiment. The straight line is that followed by birds without contact lenses. The dotted path is that of birds with contact lenses but no magnets attached to their heads.*

within it and the homing pigeon to have a magnetic compass, might it also be possible that homing pigeons have a magnet within their bodies? This question occurred to Dr. Charles Walcott of the State University of New York at Stony Brook. In order to answer it, he had to use two of the newest instruments available, the SQUID (mentioned earlier) and the electron microscope. Since the magnetic crystals were extremely tiny and could be seen *only* under the electron microscope, one first had to know where to look for them. To determine where to look with the electron microscope, Walcott used the ability of the SQUID to sense the minute magnetic field given off by the crystals. This was a difficult and time-consuming process, but he succeeded. He found the same crystals of magnetite, as a submicroscopic mass located on the surface of the pigeon's brain.

Walcott reasoned that the same simple mechanism used by bacteria could not possibly work for the pigeon; the mass of crystals was far too small to exert any push or pull on the pigeon, and some sort of information-transfer system from the magnetite crystals to the brain had to be involved. He found that the mass of crystals was full of nerve fibers that seemed to go into the brain. However, he did not know—and we still don't know today—exactly where those fibers go inside of the brain. At any rate, it appears quite certain that the magnetite crystals "tell" the pigeon's brain the exact direction of the Earth's magnetic field, and that the pigeon uses this information to navigate with its amazing precision.

Since Walcott's work, many other scientists have continued this investigation. As a result, we now have detailed information on the exquisite sensitivity of this system (it is far more sensitive than our very best magnetic compass), and we know that it is present in almost all types of living organisms—probably including the human being. Dr. Robin Baker at the University of Manchester, England, has tentatively located our "magnetic organ" in the posterior wall of the ethmoid sinus (located high up at the back of the nasal passage, just in front of the pituitary gland).

In several controversial experiments, Baker has shown that humans have an innate ability to sense the direction of magnetic north, and that this ability can be blocked by placing a bar magnet against a person's forehead for only fifteen minutes! Baker also claims that a subject's directional sense is disturbed for as long as two hours after application of the magnet. Perhaps the strong field of the bar magnet disturbs the normal orientation of the magnetite crystals, resulting in a loss of directional-sensing ability until they return to their normal orientation. If Baker's experiments can be confirmed and evidence obtained for deposits of magnetite in the ethmoid sinus, it is probable that similar nerve pathways extend from them to the brain, as have been found in other animals.

My favorite animal, the salamander, has *two* separate magnetic navigational systems. One provides a simple compass, so that when the salamander is traveling "cross country," it will go in a straight line (this is important, because it cannot go far without water). The other system enables it to return to the exact spot where it was hatched from its egg in order to mate and lay more eggs.*

---

*Back in the 1960s, I became acquainted with a biologist on the West Coast who was studying the amazing navigational ability of a certain salamander. This animal hatched in mountain streams, and after it passed through its larval stages and became an adult, it left the water. It would then wander many miles over very rough mountain terrain for a few years, and when it became sexually mature, it would find its way back to the exact spot where it had hatched—an incredible feat!

The biologist had studied the possibility that the animal visually "remembered" the terrain it had crossed over, but he found no evidence to support this idea. He was looking at the sense of smell when I wrote to him and suggested that perhaps a magnetic sense was operating in the salamanders. In his reply, he wrote, "If I cannot prove olfaction [smell] is the navigational system, I would rather give the problem to the university chaplain before I would consider a magnetic sense." So great is the power of dogma.

The magnetite-containing "magnetic organ" that is probably present in most life forms, including humans, is closely connected to the brain. It has been shown unequivocally to be a sense organ that informs the organism of the direction of the Earth's magnetic field. It is possible that it is also sensitive to, and reports information on, the micropulsation frequencies, although this aspect has not been studied. As if to emphasize the importance of the relationship between the geomagnetic field and living organisms, nature has provided us with yet another organ that also senses the field and that extracts even more significant information from it.

The Pineal Gland's Magnetic Sense

As mentioned earlier in this book, the pineal gland is a tiny, pinecone-shaped structure in the exact geometric center of the head. As you'll recall from the earlier discussion, Descartes believed that this gland was the "seat of the soul." While this is undoubtedly incorrect, the pineal has gone from the status of an insignificant curiosity to being considered the "master gland" of the body.

The pineal has an interesting history. It is the remnant of the "third eye," which was located on the top of the head in many primitive vertebrates. In this position it probably did not register images but was responsible for measuring the intensity of natural light, so that an animal could alter its coloration to better match its surroundings. While a few primitive animals (notably, the lamprey eel and the hagfish) still have a third eye that performs this function, in most life forms the pineal slowly sank from the surface to deep within the brain structure.

It has been only during the past decade that scientists have discovered how important this structure is. The pineal produces a veritable pharmacopoeia of active chemical substances. Some regulate the operations of all other glands in the body (including the pituitary, the former "master gland"); others are major neurohormones (such as melatonin, serotonin, and dopamine), which regulate the level of operations of the brain itself.

The pineal is the "clock" that the mechanists postulated was the source of biological cycles. The cyclic pattern of sleep-wakefulness is dependent upon the level of melatonin secretion by the pineal. It was first determined that a part of the output of the retina was diverted to the pineal, where it was sensed as the day-night cycle, and melatonin secretion was adjusted accordingly. More recently, it has been shown that the pineal is also sensitive to the daily cyclic pattern in the Earth's magnetic field. Melatonin secretion in human subjects may be changed at will by exposure to steady magnetic fields of the same strength as the geomagnetic field. Apparently, nature determined that biological cycle activity was too important to be left to one environmental signal alone.

At the present time there is great interest in the psychiatric community about the probability that abnormal secretion of neuro-hormones by the pineal is linked to many behavioral abnormalities. Medicine, in general, has become aware of the fact that disturbances in the biocycle pattern are of considerable clinical importance. For example, the primary effect of a chronically abnormal biocycle is the production of chronic stress syndrome, a condition that produces a wide variety of clinical problems, including a marked decline in competency of the immune system.

Most recently, it has been shown that the time when cancer chemotherapeutic drugs are administered during a patient's biocycle is a major determiner of their effect. Given at the appropriate time, they are more effective against cancer cells and produce fewer side effects than if given at the wrong time. The National Cancer Institute considers this important enough that it will conduct an in-depth study of the phenomenon.

Nature intended the pineal to simultaneously receive the same signals from the daily pattern of day-night and the same rise and fall in strength of the geomagnetic field. Obviously, when one or both signals are abnormal, the pineal does not respond in the normal fashion, and the body's biological cycles become disturbed—with important clinical consequences.

* * *

The total-body effects of external fields are mediated through at least two highly specific and sophisticated internal organs: the magnetic organ, which is composed of minute crystals of magnetite and is closely connected to the central nervous system, and the pineal gland, which is a part of the brain. The presence of one or the other of these organs in species as diverse as bacteria, insects, fish, amphibians, and mammals indicates that they are vitally important to normal life functions, and that this mechanism originated very early in evolution. The only conclusion possible is that living organisms sense the Earth's geomagnetic field and derive vital information from it.

Not very long ago, such a statement would have been viewed as rank heresy by the scientific establishment. However, the fact that living things have this capability should not come as a surprise, considering the nature of the Earth's normal electromagnetic field. If we use the concept of the electromagnetic spectrum and arrange the various components of the normal electromagnetic environment (with no man-made fields present) on a linear scale of increasing frequency, we can visualize the relationships.

It appears that, over 2 billion years of evolution, living things have taken advantage of the two portions of the electromagnetic spectrum that could be depended upon to always be present: the geomagnetic field and visible light. In this view, it is no more surprising that life developed specific organs to sense the geomagnetic field and to derive timing information from it than that it developed specific organs to sense and derive information from light.

Research on the sensitivity of the magnetic organ and the pineal has been done with DC magnetic fields, but the micropulsation frequencies have been neglected. If we expand that portion of the electromagnetic spectrum pictured in Figure 3-2 to show only the Earth's steady magnetic field and the very slowest frequencies, we find that the micropulsation frequencies have a peculiar spectrum of their own.

The frequency spectrum of the micropulsations is practically identical with that of the EEG and MEG of all organisms that are sufficiently developed to have a brain. While there is no evidence for any sensing of micropulsation frequencies, it is interesting that the major component of this frequency spectrum, 10 Hz, is also the major component of the EEG and the MEG. In addition, it is the frequency most commonly used for electrosensing by those aquatic organisms that use this mechanism to locate prey. As we shall see in later chapters, frequency may be a more important parameter than field strength for biological effect.

THE DUAL NERVOUS SYSTEM COMPLETED

Over the past twenty-five years, the discoveries described above have produced a new vision of living things, one that has returned electromagnetic energy to a position of prominence. Drawing upon the concepts of information theory and solid-state

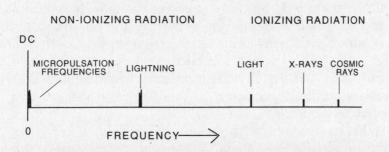

FIGURE 3-2. *The Earth's normal electromagnetic spectrum. There are two components to the normal geomagnetic field: the DC, or steady-state, field, which actually varies on a diurnal basis; and the micropulsations, which are extra-low-frequency pulsations in the field. Lightning produces frequencies in the thousands of cycles per second, but its occurrence is variable, depending upon the weather. Light is part of the electromagnetic spectrum, with frequencies in the trillions of cycles per second. Both the geomagnetic field and light are "quasistatic"—that is, they are always reliably present but display daily fluctuations in intensity. Frequencies higher than light have more energy per photon and are referred to as "ionizing" because they produce damage in cells by ionizing their components. X-rays and cosmic rays are the most common types of ionizing radiation.*

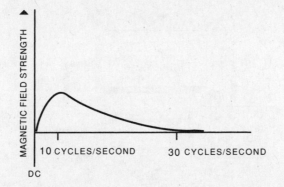

FIGURE 3-3. *The frequency spectrum of the geomagnetic field. The DC component is much stronger than the micropulsations, which stretch from just above DC to about 30 cycles per second. Note that the strength of the micropulsations is greatest between 7 and 10 cycles per second.*

physics, and aided by instruments of vastly improved sensitivity
and sophistication, we have described electronic control systems
within the body that regulate such functions as growth and healing and that also serve as the substrate for our internal control
and communications systems. Application of the same technology
to the relationship between the external energies in the Earth's
geomagnetic field and living organisms has revealed that living
things are intimately related to this field and derive vital, basic
information from it.

What has emerged is the basic outline of a dual nervous system:
a primitive analog component that appeared early in evolution, and
a more sophisticated digital, nerve-impulse component of more recent origin.

What I have described is the essence of the latest scientific
revolution, which is providing us with a greatly expanded vision of

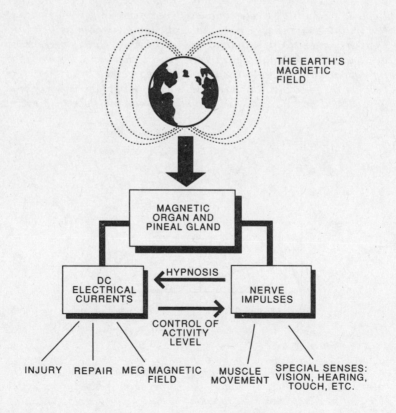

FIGURE 3-4. *Schematic outline of the dual nervous system.*

the complexity and capabilities of living things. Those embarrassing little oddities that the chemical-mechanistic theory could not explain are beginning to be understood by advocates of this new paradigm.

On the surface, it might appear that we are bringing vitalism back into biology, but that would be a mistaken view. What we are doing is bringing *energy* back into the living system and assigning it the task of organizing and controlling the total organism and its most important functions.

Medicine has come full circle, from the mysterious energies of the shaman-healer to the scientific understanding of the life energies of the body and their relationship to the energies of the environment. This scientific revolution has simultaneously enriched the concepts of technological medicine and supported the ideas of energy medicine. What is emerging is a new paradigm of life, energy, and medicine.

Part two
Electromagnetic Medicine

CHAPTER FOUR

TURNING ON THE BODY'S ELECTRICAL SYSTEM: MINIMAL-ENERGY TECHNIQUES

Research is to see what everyone else has seen and to think what nobody has thought.

ALBERT SZENT-GYORGYI

Just as orthodox science bitterly opposed the new scientific revolution in biology and biophysics, so has orthodox medicine opposed the growing interest in alternative medicine. However, these two basic medical philosophies, which have been poles apart, are beginning to move closer together. Practicing physicians are beginning to accept and use some "alternative" medical techniques. Major opposition at this time seems to be coming primarily from the medical academicians who are much more firmly wedded to the earlier dogma.

One of the most interesting developments in this area has been the gradual coalescing of many disciplines of alternative medicine into a single entity known as energy medicine. This has come about, at least in part, because of the latest scientific discoveries discussed in the preceding chapter. The basic idea is that an energy system within the body can be directly influenced, in a variety of ways, to produce a healing effect.

However, there is still more than a hint of vitalism in this popular concept, and "body energies" are vaguely defined and poorly understood by most proponents of this view. Unfortunately, many seem to prefer to retain the mystery rather than master the scientific basis that really exists for healing. In my opinion, this is as serious a problem as the rejection of the concepts of energy medicine by orthodox physicians. If orthodox medicine is to find a way out of the blind alley into which it has been led by technology, it will need the help of energy medicine. In order for this to happen, energy medicine must be placed on as firm a scientific basis as possible. In this way we may be able to resolve the pressing problems that now plague our highly developed system of medical care.

THE SCIENTIFIC BASIS FOR ELECTROMAGNETIC MEDICINE

In chapter 3, I presented the scientific evidence for the existence of electromagnetic energy systems within the body that control growth and healing, regulate the level of activity of the brain, and produce vitally important biological cycles by deriving timing information from the natural electromagnetic environment of the Earth. Most of the research was done on "lower" animals; however, I believe that the same systems are operating in all living things, including human beings. After all, we have the same basic functions of growth, healing, and biological cycles as other living things. There is no reason to believe that because we are so much "higher" on the scale of development, we should differ at this most basic level.

As soon as I discovered the overall pattern of the DC electric field in the salamander, the next question that occurred to me was whether this pattern was the same for all animals, including humans. I started lower down on the scale of evolutionary development—far below the salamander—and worked my way up. I found that all animals that had developed any kind of nervous system had DC fields that spatially matched the anatomical details of that system. The flatworm, planaria, for example, is not much to look at and is quite a bit more primitive than even the earthworm. Yet it has a collection of nerve cells near its "head," with nerve fibers running from there to the "eyes" in front of this "brain" and to the rest of the body, back to the tail. The "brain" was electrically positive, and the nerve fibers coming from it were increasingly negative the farther out they went, just as in the salamander. Also as in the salamander, chemical anesthesia produced a drop in these potentials to zero.

Going up the scale from the flatworm to the salamander produced the same results. In each type of animal tested, the details of the internal DC electrical field always mirrored the design of the nervous system, with the areas where nerve cells were concentrated being positive and the nerve fibers radiating out from these areas being negative.*

THE HUMAN ELECTRICAL CONTROL SYSTEM

After surveying animals that are lower on the evolutionary scale than the salamander and finding the same internal DC system in each despite major differences in anatomy and physiology, I turned to the human being. It was not possible to use the same measurement techniques on a person as on a salamander. Moving from the salamander to the human being was like moving from the human being to the elephant; the size differential required major new designs in electrodes. Fortunately, at that time I was working with a talented physicist, Dr. Charles Bachman of Syracuse University. He and I talked the measurement problem over, and a few days later he appeared with an odd-looking object that hardly resembled an electrode, but that worked like a charm.

We found the same DC system pattern in the human being as in the salamander. This was not unexpected because, as I noted before, the salamander is the basic vertebrate, with all the same parts and arrangements of structures—including the nervous system—that humans have: the same concentrations of nerve cells in the brain; the cervical enlargement of the spinal cord (although in humans, these nerve cells are in the neck rather than in the shoulder region); and the lumbar enlargement. These were all found to be positive in the human being, with the nerve-fiber outflows from each increasingly negative the farther out we placed our measuring electrode.

We found the same relationship between the level of conscious-

*It was impossible for me to measure any potentials from the most primitive single-cell animals. However, I did study one very odd single-cell animal, Physareum, an amoeba that lives as a single cell in moist places on the ground. Occasionally, many of these solitary amoebae come together into a single mass that moves as a single unit, seemingly with a "mind" of its own. This mass aggregate of single cells moves across the ground surprisingly fast and in one direction. There is no organization whatsoever at this stage; all the cells look the same, and any one separated from the mass could survive normally as a single cell. My electrical measurements showed that the "head" end was always negative. Somehow, all the independent, single cells that made up the mass generated a "normal" DC electrical field.

ness and the strength of these DC currents. Sleep produced a modest drop, and deep general anesthesia resulted in a drop in the potentials to zero in the same sequence, starting with the brain. At the time that we were doing these experiments, a number of physicians—mainly in Eastern European countries—were experimenting with electrical currents to produce anesthesia and were claiming good results. Their techniques, however, were not like those we had used on the salamander, and I was not convinced that this procedure was entirely safe for human beings. But I felt that if our experiments on magnetic anesthesia in the salamander could be duplicated in people, this type of anesthesia might be safer than the standard chemical anesthesia. We did apply (although with tongue in cheek) for the funds to get an electromagnet that was large enough for the technique to be tried on humans. As expected, we were turned down.

It also was impossible to use the Hall effect to determine whether the DC currents flowing in the human arm were semiconducting and to show that magnetic fields influenced these currents. However, we *were* able to show a relationship between the Earth's geomagnetic field and human behavior.

I started with the premise that if my theory were correct, there should be some behavioral effect on living organisms when the normal geomagnetic field was disturbed by magnetic storms. (The entire relationship between living things and the Earth's normal geomagnetic field is the subject of chapter 7, in which magnetic storms are explained.) I theorized that in human beings this effect might be most evident in people suffering from true psychoses, such as schizophrenia or manic-depressive psychosis. This was in the early 1960s, before the discovery of tranquilizers and other psychotropic drugs; at that time, severely disturbed patients were simply hospitalized.

I began to collect data on the daily hospital admission rates of such patients, and I related this incidence to the occurrence of magnetic storms. After a few months, my rather odd behavior came to the attention of one of my colleagues, Dr. Howard Friedman, chief of the Psychology Service at the VA Hospital in Syracuse, New York. Rather than believing that *I* needed treatment, he was fascinated. He turned out to be a valuable partner in my research from that time on, as he was well trained not only in psychology but also in statistics and experimental design. Together, we set up a retrospective epidemiological study. Such a study looks at what happened in the past and attempts to show relationships between the incidence of a disease process and some other variable. In this case, we looked back at several years of admissions of these types of patients to state mental hospitals, and we compared these numbers with the incidence of magnetic storms.

We found a significant relationship between the rate of admissions for people with schizophrenia and manic-depressive psychosis and the occurrence of major magnetic storms. As could be expected, this produced a "storm" of another type—widespread criticism that such a relationship was not possible. In response, Friedman devised another experiment. He set up a scale of criteria for "ward behavior" of hospitalized psychiatric patients, ranging from "very disturbed" to "normal," with about ten steps between the two. Each day the nurses on the psychiatric wards filled out this form for each patient. Later, we related the numerical scores to the level of neutron flux measured at a laboratory in Canada. We chose this parameter because it accurately reflected the state of the Earth's magnetic field. Again, a positive correlation was found.

Somehow, patients with severe mental illness sensed the state of the Earth's geomagnetic field and responded to disturbances in it with increased behavioral abnormalities. Could a similar situation be present in other human beings as well, but expressed in a more subtle manner? Unfortunately, we did not have the necessary funds to begin such a project.

We concluded from these studies that human beings possess the same internal DC electrical control systems as other animals, and that through these systems they are behaviorally related to the Earth's geomagnetic field.

LINKING HYPNOSIS TO THE ELECTRICAL SYSTEM

In addition to his other talents, Howard Friedman was a talented hypnotist. Since we were both interested in how the brain works, hypnosis seemed to be a good place to start. While I was aware of the stories of how one could "hypnotize" a chicken, I believed that hypnosis was primarily a *human* affair, but one that was surrounded with controversy. Having been born in Mesmer's experiments on "animal magnetism" and first called "mesmerism," it had been tainted from the start with charlatanism. Even today, there are prominent psychologists who insist that hypnosis simply does not exist, that it is nothing more than an intense desire on the part of the subject to please the hypnotist. Because hypnosis appeared to be a rather odd state of consciousness, and because I had a good way to measure consciousness with DC potentials, we decided to do a little experiment in this "fringe area" of science.

What we discovered was much more important than we had expected. We found that we could reliably determine whether a subject was truly hypnotized or was simply trying to please Dr. Friedman. In true hypnosis, the DC potential from the front to the back of the head (which is actually a measure of the brain's midline DC current) undergoes a drop in strength similar to the drop that occurs during very deep sleep. If the subject was only trying to please us, he or she was mentally active, and the DC potential went *up* in strength. We therefore concluded that hypnosis is real, that it has a measurable electrical correlate, and that it represents some change in the subject's level of consciousness.

What is interesting about the technique is that the hypnotist can still make contact with the subject's conscious brain and give it orders, or "suggestions," that the subject will carry out, either at that time or at some time in the future (this is known as posthypnotic suggestion). Therefore, hypnosis is quite different from other types of changes in consciousness level.

One of the most interesting aspects of hypnosis is its ability to produce anesthesia. It is possible to give a truly hypnotized patient the suggestion that a part of the body is numb, cold, and unfeeling, and to then perform minor surgery on that part without the patient's perceiving pain.*

We had already shown that general anesthesia in human patients is produced by a fall in the normal DC electrical current across

---

*Major surgery may also be done under hypnosis; however, the patient must be prepared for the surgery with multiple episodes of hypnosis and testing. When I was a medical student, hyperthyroidism was a great problem; there was no effective medical treatment for this condition, and surgery was extremely dangerous. Under anesthesia, a hyperthyroid patient had extremely variable blood pressure, often going from dangerously high levels to total collapse in seconds. Anesthesia deaths during surgery were not uncommon.

I was privileged to watch a thyroidectomy for hypertension that was done using hypnosis alone as the agent of anesthesia. The patient was fully awake and relaxed. I was very impressed by her ability to control bleeding for the surgeon: he simply told her that he had a troublesome "bleeder" in a certain section of the wound, and she was able to stop the flow of blood within seconds. In this case, the patient had been prepared by the hypnotist for six months prior to the actual surgery. During these sessions she had been "taught" to totally anesthetize the area and to control both blood pressure and bleeding. Obviously, this is a time-consuming procedure, and it is little used today.

the brain, which then seems to produce similar declines in the DC potentials in the remainder of the body. We had also shown that local anesthesia, produced by blocking the nerve to a single part of the body, results in a drop of the DC current to zero in that area alone. We concluded that the perception of pain is directly controlled by the status of the DC currents, either in the entire body or in any local area.

The theory that we developed was that if the local anesthesia produced by hypnosis is real, it should be accompanied by a similar drop in the DC current in the anesthetized area. If this occurred, its meaning would be far more significant than substantiating that the anesthesia of hypnosis was real. *It would mean that the conscious mind, under hypnosis, could control the level of activity of the DC control system.* The implications of this for energy medicine would be enormous.

For our next experiment, we used only our best subjects—that is, those who could easily be put into deep "trance" states. We instructed them that their left arms and hands were numb and cold. For each subject, we continuously measured the DC across the brain and, separately, along the left and right arms from the beginning of hypnosis. We found that the DC across the brain declined as expected, indicating a true hypnotic state. The DC along the left arm began to drop as the suggestion of numbness and cold was given, and after a few minutes it reached zero. At that time, the subject did not respond to a pinprick on the left hand or lower arm. There did appear to be some "crossover" in that the DC along the right arm fell a modest amount along with the left, but the subject still sensed the painful stimulus on the right side. The subject was then given the suggestion that feeling in the left arm was returning; over a few minutes, the DC potential in that arm returned to normal. The decline in the DC potentials along the anesthetized left arm was exactly the same as that seen during standard chemical nerve block.

Under hypnosis, humans may be given verbal commands to the conscious digital-system portions of the brain, which can then effectively control the operations of the DC analog system. Since the primitive analog system controls growth and healing, it is possible that under certain circumstances, conscious thought can cause healing.

This discovery means that a link exists between the conscious brain and the primitive analog brain whereby the latter can be controlled by the former. It seems reasonable to postulate that this linkage can be exploited by some of the techniques of energy medicine—namely, visualization, biofeedback, meditation, and possibly any technique that produces a profound belief in the success of the treatment, such as the placebo effect.

A scientific basis thus exists for the concept of energy medicine. The internal DC electrical control system does exist in humans as in all other animals. It may be accessed by a variety of techniques, and desirable changes in its activity may be produced to effect healing. There is yet more scientific evidence for this theory; that evidence will be discussed in relation to specific techniques of energy medicine.

THE THREE TYPES OF ENERGY MEDICINE

The scientific basis for energy medicine is often poorly understood, and body energies are still considered by many practitioners to be mysterious, unknowable entities. It is important that this reintroduction of energy back into medicine not mean a return to the world-view of the shaman-healer of preliterate times, with the invocation of mystical forces. Nor should it come to mean the uncritical application of electricity and magnetism, the two physical forces thus far identified as operating in the body. If energy medicine is to assume its rightful place as an effective form of medical therapy, it must be based on established scientific principles and careful scientific experimentation.

Proponents of many different alternative therapeutic techniques have eagerly adopted bodily electromagnetic forces as their own, typically with only a foggy notion of what has been scientifically established. In some cases, extravagant claims have been made, and advocates of each specific therapy have adopted the position that theirs is *the* true energy medicine. This is a most unfortunate situation, because these techniques often produce significant effects and are capable of doing harm if used improperly. Unless we investigate each of the various therapies from a scientific point of view and apply them in a rational, safe fashion, the entire field of energy medicine is in danger of being consigned to the dustbin of medical history. However, such a massive investigation will not be easy: our present scientific knowledge of the operations of these forces is incomplete, and in addition we are faced with a bewildering array of therapies, ranging from acupuncture to Zoroastrianism.

On the basis that the valid techniques of energy medicine act through the internal energetic control systems of the body, we can classify the various techniques into three broad types, depending upon the energy levels used. This must be viewed only as a preliminary attempt at classification, for, as we shall soon see, not every technique can be accurately categorized in this way.

First, there are techniques in which no external energy is administered to the body (examples are hypnosis and visualization), and in which the treatment methods attempt only to activate preexisting energetic control systems. I refer to these as *minimal-energy techniques*.

The second category encompasses techniques in which external energies are administered to the body, but in amounts similar to those that the body itself uses in its energetic control systems. These include acupuncture and, probably, homeopathy. Since such techniques can be seen as reinforcing an inactive or inadequate energetic control system, I call them *energy-reinforcement techniques*.

In the third category are techniques in which energy is administered to the body in amounts greater than those that occur naturally. These are obviously related to the general practices of technological medicine, in the sense that the normal system is replaced by this externally derived energy. I call these *high-energy transfer techniques*.

To illustrate the application of this classification scheme, I'll apply it to three of the oldest medical techniques: the shaman-healer, acupuncture, and the local application of a lodestone (a naturally occurring magnetic mineral).

This theoretical scheme is in its infancy, and we are left to make do with indirect or fragmentary evidence in our efforts to understand how energy medicine works. However, the common thread that runs through our evaluations of the techniques is that of the energy systems of the body. When considering each technique, we have to determine whether it acts by influencing these energy systems. Because the energy systems thus far identified are based on electromagnetic energy, the technique under consideration has to have some ability to change the operation of these energies within the body.

While we do not yet know all there is to know about electricity and magnetism, it would be foolhardy to base our concepts on the supposed actions of other forces about which we know nothing. To do so would mean reintroducing mysticism back into energy medicine—and this must be avoided if we wish to gain any recognition for our new view. If we can provide energy medicine with respectability at this crucial time, we can clear the way for future develop-

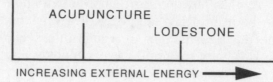

FIGURE 4-1. *On a spectrum of increasing energy application starting from zero, the shaman-healer applies no energy, simple acupuncture needles introduce very small amounts of electricity, and lodestones have a local magnetic field considerably greater in strength than the Earth's. The shaman-healer (provided he does not administer any natural herbs or other medicinals) simply activates the patient's own internal DC system. Acupuncture (the nonelectrical type) applies a very modest amount of external electrical energy to specific portions of the internal DC system, and it may be classified as an energy-reinforcement technique. The lodestone exposes the portion of the body to which it is applied to a much higher magnetic field than normal, and it may therefore be classified as a high-energy transfer therapy.*

ments that will expand the horizons of this field beyond anything we can now conceive. Perhaps new physical forces, or unexpected variations of those that are already accepted, may be found through deeper explorations of the energy mechanisms of living things.

In the remainder of this chapter, I will discuss a few of the minimal-energy techniques; the other two classifications will be explored in the next two chapters. I will discuss only those therapies with which I have had personal experience and to which I can justifiably ascribe either a theoretical or an experimentally proven mechanism of action. I have undoubtedly omitted a number of useful techniques; for this I offer my apologies to the practitioners of these techniques. I have also deliberately not mentioned a few popular "therapies" for which I can find no scientific basis.

MINIMAL-ENERGY TECHNIQUES: THE USE OF THE MIND

The minimal-energy techniques are directly based upon the internal energy systems we have been exploring, and it seems appropriate to continue discussing them in this context. The list of minimal techniques is long, and it includes meditation, biofeedback, visualization, hypnosis, suggestion, placebo, religious experience, and religious healing. The one aspect common to all is that the conscious mind is being influenced in a desired direction, with the body then following the dictates of the mind (the phenomenon of minimal-energy techniques has also been called self-regulation). With these techniques, we are actually dealing with the basic mind/body problem, but at a practical rather than a theoretical level.

To orthodox physiology, this is nonsense. First of all, there are no known connections between the conscious mind—presumably residing in the digital nervous system—and that portion of the digital system known to control those more primitive automatic systems that regulate blood pressure, blood distribution, body temperature, respiration, and digestion. These functions are the province of the autonomic nervous system, so named because it functions automatically, without intervention by the conscious mind. For example, it is not necessary for us to consciously regulate respiration or heart rate. As to the more primitive functions not controlled by the autonomic system, such as growth control and biological cycles, orthodox scientists simply reject the idea that the mind can have any direct effect on them.

Controlling the Autonomic Nervous System with the Conscious Mind

In many of the minimalist techniques, the first step in therapy is to consciously produce an increase in the blood flow to the hands. Biofeedback practitioners accomplish this by instrumenting the patient's hand with a thermocouple and instructing the patient to make the recorder pen move up on the paper. In hypnosis, the subject is simply instructed to make his hands warmer by increasing the local blood flow. These techniques work reliably in the majority of subjects, and the advantage of this simple demonstration is that it convincingly shows patients that they *can* control their own bodies. Exactly how this control is exerted is not yet known. Either there is a direct connection between the conscious mind and the autonomic system, or the DC system is involved, as we showed with local anesthesia.

The most remarkable demonstration of the ability to control blood flow to a particular body part that I have witnessed occurred in the course of a joke played on me by a patient and Dr. Friedman, my hypnotist colleague. As the Vietnam war was winding down, I had a patient who had been injured in action. He had been left with a persistent, severely infected nonunion of a fracture of the tibia (the shinbone). Unfortunately, like so many others in Vietnam, he had been addicted to hard drugs, a habit he had broken with great difficulty. Because his treatment required several episodes of surgery, he was deathly afraid that the use of any postoperative narcotic for pain would trigger his addiction again. I sent him to Dr. Friedman, who found him to be an easily hypnotizable subject and instructed him in the basics of self-hypnosis. With this simple technique this young man was able to control the postoperative pain from his first surgery.

He progressed well, and we finally came to the last surgery. This would consist of placing a bone graft in the defect in his tibia. Because the wound was left open, it was necessary to completely cover the graft with a blood clot at all times after the surgery. I told the patient that if the blood clot appeared to liquefy and not cover the bone graft, I would simply take some blood from a vein in his arm and squirt it into the wound.

On the fourth postoperative day, the clot seemed to be shrinking. "If it's any worse tomorrow, I'll put some more blood in it," I told him—to which he replied, "Don't worry, Doc. I'll be okay tomorrow." Sure enough, when I removed the dressing the next day, there was so much blood that it was oozing out of the wound. I must have looked puzzled, because he laughed and said that he and Dr. Friedman had cooked up a little scheme. Dr. Friedman had taught him in one session to use self-hypnosis to increase the blood flow to that part of his leg, cautioning him not to overdo it. Early that morning he had gone into his self-hypnotic state, but he had overdone it a bit!

Clearly, the conscious mind can control the autonomic nervous system with exquisite precision, through either a direct link or the DC system. I have often used hypnosis and biofeedback therapies for patients with hypertension and chronic pain, and I believe this approach to be not only as effective as the standard drug therapy but also a great deal safer.

The masters of autonomic-nervous-system control are probably the Indian yogis, who are able to control their intestinal movements, slow their respiration and heart rates to almost zero, raise or lower their body temperatures, and accomplish other "miracles" prohibited by modern science. Perhaps one day we'll be able to conduct a full scientific study of these remarkable people.

After a patient has acquired the hand-warming ability, a variety of other minimal-energy techniques may be used for more significant purposes, such as growth control.

Controlling the Immune System with the Conscious Mind

Over the past decade, it has become evident that a similar relationship exists between the conscious mind and the immune system. This is now an accepted reality in orthodox medicine, and it has even become a scientific discipline (with the jawbreaking name of "psychoneuroimmunology"). The only problem is that there are, as yet, no known digital neural links between the brain and the various component parts of the immune system, although at the cellular level there appear to be definite relationships between the nerve cells and the immune-system cells. This comes as a surprise in view of the fact that the two cell lines diverge very early in embryogenesis.

The relationship between the central nervous system and the immune system is complex. There are studies that have shown a relationship been mood states, such as depression, and the level of competency of the immune system. But despite a great deal of research, we still do not understand this completely. While the experiments on hypnotic anesthesia discussed earlier in this chapter may provide links between the brain and the immune system via the DC system, this possibility has yet to be explored.

HOW MINIMAL-ENERGY TECHNIQUES WORK

I believe that the link between the conscious mind and the DC system is basic to the minimal techniques of energy medicine. While such a link was demonstrated in our experiments on hypnotic anesthesia, this does not mean that the link, or gateway, is opened only by this type of classical hypnosis. Self-hypnosis (or autohypnosis) can also be used. This technique is initially taught to the patient while he or she is under a classical hypnotic "trance." Even after it has been mastered by the patient, self-hypnosis still requires either such a trance or any overt change in the state of consciousness. Concentration and belief on the part of the patient are the major components necessary. My basic premise is that whenever a patient develops a profound belief in the efficacy of *any* treatment, he or she is in a state resembling self-hypnosis and has access to the DC electrical control systems.

For example, consider the clinical situation of "hysterical anesthesia," in which a portion of the body is declared by the patient to be completely lacking in feeling, or "dead." This is primarily a psychological condition, and the body part is generally one that has been involved in a situation that has brought the patient severe mental anguish—for example, if the patient's arm or hand has struck and injured someone. The anesthesia is "real" in the sense that the patient does not respond to painful stimuli in that area. While the patient is not hypnotized to produce the effect, the two conditions are quite similar in that both the hysterical patient and the hypnotized subject may be considered to have a profound belief that the arm is numb.

Anecdotal and objective evidence exists for cures of cancers produced by the minimal-energy, or self-regulatory, techniques. In physics, a body in motion or a system in a state of operation will continue doing the same thing until an outside force acts upon it to change or "perturb" it. In biology, the growth of a tumor cannot be stopped simply by a wish that it would do so. Therefore, the cures resulting from these therapies must involve the action of some real force that perturbs the malignant process. If we are to exclude the operation of some mysterious, unknowable force (as I believe we must), we are left with the electromagnetic force as the only viable candidate.

The evidence I have presented so far indicates that the conscious mind, in the state of profound belief, can control the operation of the body's DC electrical growth-control system and of the immune system to produce the results that have been observed. Given the present state of knowledge in this area, this theory at least provides a valid, scientific framework within which to proceed with further investigation.

Profound Belief and the Placebo Effect

The best example of the operation of profound belief is the placebo effect. A famous case is that of a man who was dying of inoperable cancer during the time of the laetrile controversy in the 1960s, prior to the availability of chemotherapeutic drugs for cancer treatment. Laetrile is a simple chemical that was claimed to have a major anticancer property. It was promoted by some physicians who had excellent credentials, and it enjoyed wide usage for a while, although with mixed clinical results. Some patients reported complete "cures," while others experienced no effects at all.

The patient mentioned above had lost a great deal of weight and was in constant pain, and he asked his physician if he should try laetrile. The doctor saw no reason to stop him, and he told the patient that he knew of several people whose conditions had improved markedly with laetrile treatments. After beginning this treatment, the patient experienced a remarkable recovery: the tumor diminished greatly in size; the patient gained weight, and his pain disappeared.

A few months later, when the cancer was no longer palpable and the patient felt that he was making a complete recovery, he read the press reports saying that the AMA had declared laetrile "worthless." Very shortly thereafter, all of his previous physical symptoms and signs recurred, and the tumor reappeared. The patient's physician then told him that he had obtained a much more potent preparation of laetrile that *did* work, and he proceeded to give him daily injections of what was actually just distilled water. Again, the patient went through a complete remission, with the tumor size diminishing to a nonpalpable state. The patient again experienced a rapid gain in weight and a renewed sense of well-being. But a short time later, he read that after careful testing, the FDA now completely supported the AMA's opinion that laetrile was worthless. The tumor underwent a massive increase in size, and the man died within a few weeks.

There is no doubt that laetrile is basically a worthless chemical that has absolutely no specific antitumor effect. The above patient's two episodes of recovery can only be ascribed to his profound belief, reinforced by the authority of his physician, that the treatments would work. Somehow, under this mind-set, the conscious mind gains access to the DC electrical control system, producing control over the growth of the tumor.

The placebo effect is a basic and extremely important tool that traditional medicine has not only discarded but has actually legislated against. The case just discussed could not happen now, because the physician would be in real danger of losing his license.

When I was in medical school, before antibiotics were available, we were taught that *how* we treated patients was as important as what we treated them with. Because few really effective medications were available, it was necessary to make use of patients' own abilities to resist disease. In those days, this was often referred to as the proper "bedside manner" or "doctor-patient relationship." But I always felt that this trivialized a most important part of medical practice. If a doctor is really interested in the patient and is able to develop in the patient a conviction that he or she will get well with the doctor's help, real physiological changes occur within the patient's body that actually assist healing. The doctor-patient relationship thus has to do not only with the patient's mind, but with the entire body as well.

I realized long ago that even in such a mechanistic discipline as surgery, the proper relationship between the surgeon and the patient is as important as surgical skill. I once worked with a surgeon who had acquired a high level of skill in the mechanics of surgery. At that time, the most demanding surgical procedure done was the thoraco-lumbar sympathectomy for hypertension. This entailed the complete removal of the sympathetic portion of the autonomic nervous system, a continuous chain of nerve fibers and ganglia running from the upper portion of the neck through the chest and abdomen and deep into the pelvis. This surgeon was able to remove this entire anatomical structure in one piece through one small incision along the lower ribs, in an average time of thirty minutes. Our operating room was often crowded with medical students and junior residents who came simply to see this wizard operate. His skill at other surgical procedures was equally impressive. However, regardless of the surgical procedure done, his patients had more postoperative complications and deaths than patients operated on by any other surgeon at the hospital.

There was another surgeon on the staff who had almost the same technical competency. He was able to do the same procedures, but he took a little longer, and he lacked the panache of the "hero surgeon." His postoperative complications were half those of the other surgeon. The only difference I could see between the two men was their relationship to their patients. The master technician never spoke to his patients; he grunted at them for all of sixty seconds on the day prior to surgery, and he left it to the residents to answer patients' questions and calm their fears. The other surgeon spent as long as necessary to explain in detail what he was going to do and to answer patients' questions, leaving them with a feeling of confidence in their surgeon.

The term "placebo" may thus be expanded from the textbook definition of an inactive or ineffective chemical agent, such as a "sugar pill," and applied to the *state of the patient's mind*. Taking this one step further, I believe that under the circumstance of such a profound belief, produced by *any* technique, the conscious mind of the patient is able to access the DC system of the body and produce healing. This makes it possible to rationally explain the useful clinical effects of many diverse techniques classified under energy medicine.

New drugs are tested in a double-blind manner so that neither the dispensing doctor nor the patient knows whether what is being dispensed is an effective medication or a placebo. The basic assumption underlying this idea is that the only valid treatment is one that has been clearly demonstrated to influence the "target" chemical system without the intrusion of psychological effects. On the face of it, this seems like a very responsible and logical way to go about the business of medicine. Certainly, we don't want to have a lot of worthless drugs on the market.

However, we are missing the mark in this case. The whole aim and justification of medicine is to produce the relief of symptoms or to effect the cure of a disease. Standing in the shadows, beyond the light of present-day science, is the placebo effect, which is capable of producing the desired medical effect in sixty percent of clinical cases overall! If the placebo effect could be patented and bottled, it would make millions of dollars for the pharmaceutical company that owned it.

Perhaps the most disastrous result of the concerted attack on the placebo effect is the accompanying concept that before any treatment can be legally used on human patients, its mechanism of action must be known. Again, this appears at first to be a laudable idea—after all, we don't want to be treated with any medication or technique without knowing how it works. The problem with this idea is that the way it works must be in accord with the present chemical-mechanistic model: any technique, no matter how clinically effective in relieving symptoms or producing a cure, must be seen to work chemically or must not be used at all. This concept, rooted in the tenets of scientific medicine of the 1930s, was what finally threw the placebo effect out the window. Because there was no explanation for the placebo effect, according to the science of the day the effect simply did not exist.

Things have changed. The new scientific revolution has provided medicine with evidence for the *reality* of the placebo effect. All of the techniques listed in the beginning of this section will work, provided that the patient has a profound belief in the treatment's efficacy. In this way, the immune system is activated, the DC control system is accessed, and tissue healing is accomplished.

The minimal techniques of energy medicine are quite different from the placebo effect as depicted and condemned by orthodox medicine. In that depiction, an unscrupulous physician deceives a patient by prescribing a medication and stating that it will work, while knowing that it is actually ineffective. The entire success of this type of placebo effect rests upon the physician as a figure of total authority, with the patient surrendering his or her will to the doctor. Such patients may passively develop a profound belief in the efficacy of the medication, and some may experience a cure. However, these patients are never directly involved in their treatment in any way.

In addition, the orthodox physician often unconsciously uses a negative placebo effect, to the detriment of the patient. It is now considered to be good practice to tell all patients the "truth," and to tell those with cancer that is considered impossible to treat that they have the disease and that nothing can be done. They are not told that occasionally similar cases experience "spontaneous" regressions of the cancer, or that there are alternative therapies that occasionally seem to work. In this way, all hope is taken away, and the patient is left defenseless against the disease. Invariably, patients want to know how much longer they may expect to live. Instead of saying that it really isn't possible to guess because each patient is different, the doctor often gives a specific time frame, such as "six months." I have personally known of far too many patients who have died *exactly* on the day that a physician's estimate predicted. The authoritarian statement from the doctor produced a mind-set in the patient that not only brought about this physiologically and statistically unlikely event, but also destroyed the patient's enjoyment of the time remaining.

The ethical practitioners of minimal-energy techniques do not deceive their patients in this fashion. Instead, they tell patients from the start that they are going to cure *themselves* by learning techniques that will give them control over their own bodies. These practitioners are more like teachers than healers, guiding their patients into deeper self-knowledge and ability. This type of therapy must bring into play much more of the body's inherent control systems than the simple authoritarian placebo, and from that point of view it must be a more efficient technique. The patients are in control of their own destinies. They make their own decisions and regulate their own healing, without being dependent on an authority figure. This may be a significant point in the success of these techniques, because developing this sense of independence, a feeling of control and self-determination, has been shown to significantly reduce the stress of a serious or life-threatening illness.

Several very well-done experimental studies have clearly illustrated and substantiated this point. In 1979, doctors L. S. Sklar and H. Anisman set up two cages of rats that had been inoculated with a malignant tumor. Both cages were set up so that a lengthy electrical shock was given via a grid on the floor at random intervals, delivering a "foot shock" as a stressor. One cage had a switch that the rats could use to shut off the shock, a so-called "escape routine." Although both cages received exactly the same number and extent of shocks, the rats in the first cage were able to limit the length of time the shock was delivered, while those in the second cage had to passively accept the same amount of shock. So, one could reasonably say that the rats in the cage with the escape switch had a measure of control over this undesirable event. The rats in the second cage, who had no such control, demonstrated a significantly higher incidence of clinical cancers, a significantly higher mortality rate, and a swifter course of death.

This study was repeated by Dr. M. A. Visintainer and his colleagues at the University of Pennsylvania in 1982, except that in this case the type of cancer used was one that rats could reject with their own immune systems. It was found that the rats with escapable shock had a significantly higher incidence of cancer rejections than the rats with the same but inescapable shock. These studies clearly indicate that in any stressful circumstance, any measure of control lessens the extent of stress and results in a significant improvement in resistance factors and healthy outcome.

This teaches us several extremely important lessons. First, stress enhances and accelerates cancer growth. Second, providing patients with the maximum possible level of control over their treatment markedly reduces the stress of serious illness and improves the clinical outcome. We are rediscovering what Hippocrates discovered so long ago: to a large extent, the psychological response to a specific disease determines its clinical outcome.

I am reminded of two cases that came to my attention more than a decade ago. Two men, both about the same age, both married and the fathers of children, developed leukemia practically simultaneously. One of the men passively accepted his condition, putting his fate completely in the hands of his physicians. He endured massive chemotherapy without complaint, maintaining the attitude that the doctors knew best. The other patient completely rejected the entire situation, refused chemotherapy, and threw dishes and glassware against the wall at home. Both patients had sought help from the same physicians, who prevailed on the first patient to "counsel" the second to accept chemotherapy. But his efforts were of no avail. The first patient responded poorly to the chemotherapy and was finally given experimental drugs, which didn't help; he died within two years of the onset of the cancer. The second patient is alive and well today.

New Light on Visualization

Visualization is a technique in which a patient is instructed to "look inward" into his or her body, to "visualize" the disease that is there, and then to visualize the body's defense systems attacking the diseased tissues. The technique is used most frequently in cases of cancer in which fairly accurate localization of the diseased tissues can be made through X-rays or other imaging techniques. The patient is given the maximum information about the illness and receives complete access to laboratory data, such as X-rays, as an aid in forming a mental image of the lesion. This step certainly demystifies the disease and makes the patient part of the curative process, providing him or her with a feeling of being in control. But this cannot be all there is to it.

There are enough documented cases indicating that some patients, particularly children, can successfully form an accurate picture of the lesion itself. An excellent example is the story of a patient named Garrett Porter, as told by Dr. Patricia Norris in her book *I Choose Life*. At the age of nine, Garrett was diagnosed as having an astrocytoma, an inoperable and fatal brain tumor. He was treated with X-rays, and he began to learn biofeedback techniques almost simultaneously. After he had quickly mastered control of his autonomic nervous system, Garrett began visualization therapy with Dr. Norris. Not only did he do well clinically, but he was also able to visualize the tumor in detail. One evening, slightly more than a year after he had begun therapy, Garrett called his father and told him that he couldn't "find" the tumor anymore; in its place he "saw" a small white spot. No one except Garrett believed this. For fear of the

effect that bad news might have on the boy, a CAT scan was put off
for a month. But when it was finally done, it was clear that the tumor
was, indeed, gone; in its place was a small calcium deposit—the
white spot Garrett had "seen."

In some way, Garrett was able to actually see in his mind's eye
an accurate representation of the interior of his body. While he
obviously believed in his treatment, such belief was required only to
gain access to his internal control systems. Once this connection was
established, things happened that we cannot now explain. Simple
belief would appear to be insufficient to explain the reality of his
visualization. (Incidentally, Garrett Porter is alive today and in his
mid-teens; with the exception of some neurological defects caused
by the X-ray therapy, he is fine.)

Science "explains" the visual system in simplistic terms that
are quite inaccurate. Photons of light are focused on the retina,
where they form an image of the external world. The "pattern" of
this image is transferred to impulses in the optic nerves, and it is
ultimately expressed as a similar pattern on the optic cortex at the
back of the brain. This process is much like the process a computer
uses to form a graphic image on the screen. This graphic image is
expressed as a specific collection of individual dots; the more dots
there are and the closer together they are, the more realistic the
image will be. We call this "image digitizing."

If things were this simple, we should be able to electrically
stimulate the optic cortex with a pattern of electrodes arranged, for
example, in the shape of the letter A, and the subject would report
seeing the visual image of the letter A. The fact of the matter is that
even the letter A cannot be perceived in this way; the patient merely
reports the sensation of light. We expect that somewhere in the
brain such a patterning is expressed—on the "computer screen of
the mind," for example. But we have not found the screen. This may
be because it exists only in the state of consciousness, not in the
physical brain.

I know of one experimental circumstance in which simple images can be impressed on the conscious mind without going through
the visual system at all. This technique was discovered by Dr. Elizabeth Rauscher, a physicist, and William Van Bise, an engineer. It
uses magnetic fields generated by two coils of wire, each pulsing at
a slightly different frequency and directed so as to intersect at the
head of the subject.

When two beams of electromagnetic energy with different frequencies intersect at some point in space, a third frequency is formed. This frequency is the difference between the two original frequencies and is called either the "difference" or the "beat" frequency. For example, if one beam had a frequency of 100 kHz and the other a frequency of 99.99 kHz, the difference would be 0.01 kHz, or 10 Hz. This is a useful technique for producing extremely low frequencies (ELF) within a small volume located a distance away from the original transmitters. In Van Bise and Rauscher's experiment, the different frequencies were always in the ELF range. The size of the volume in which this occurs depends upon the diameter of the two beams. (This technique will be discussed further in a later chapter.)

Van Bise and Rauscher's blindfolded subjects "saw" such simple figures as circles, ellipses, or triangles, which could be changed by altering the frequency of one of the coils. The coils were several feet from the subject's head. The magnetic field strength from the coils was so small that no electrical currents could have been generated in the brain, and the control electronics were located in another room.

This technique seems to bypass the entire visual system as we know it, acting directly at the interface between the organic brain and the conscious perception of visual images. If this supposition is even partially correct, it is highly unlikely that the conscious perception level is based upon a digital or nerve-impulse system. The technique could not possibly have stimulated nerve impulses in any part of the brain, because the field strengths were far too low and the ELF frequency cannot produce such stimulation. It is probable that this sensitivity to such low-strength, low-frequency magnetic fields requires the presence of semiconducting DC elements. If Garrett Porter was accessing his internal DC control system, did he also gain access to his total body *morphogenetic field*, "visually" represented within his internal DC control system? And was he in this way able to accurately "see" the tumor itself? Only time and much more research will tell.

UNRAVELING THE HEALER PHENOMENON

Before we discuss the other classifications of energy-medicine techniques, we need to discuss one that cannot at this time be assigned to any of our arbitrary groups of energy medicine. The healer phenomenon is probably the oldest medical technique, with its origin in preliterate hunter-gatherer societies. Yet today it remains the least understood and the most controversial of all the techniques of energy medicine.

If the healer phenomenon is real, we do not know whether it operates via the placebo effect or whether any energy flow is involved, either from the healer to the patient or vice versa. Part of the problem lies in the deeply ingrained rejection of the entire concept by both scientific medicine and the physical and biological sciences. Nothing will produce more of an uproar at a medical meeting than asking, "What about the healer phenomenon?"

If the healer actually evokes the placebo effect, we should be able to find evidence for this in the common practices of well-known healers. The placebo effect can be activated only by two distinctly different techniques: the *teacher,* used by practitioners of minimalist techniques of energy medicine, and the *authoritarian,* practiced by physicians using known ineffective drugs.

In my contacts with healers over the past twenty years, I have found that the genuine ones adopt a businesslike attitude toward their practice. Each patient is treated, not taught. No true healers present themselves as authoritarian figures or as persons possessed of mystical powers. If the placebo effect is operating, it must be on the basis of preconditioned patients, not on any actions of the healer. While there are undoubtedly some such preconditioned patients, I have found them to be in the minority. Finally, Chinese healers also work quite successfully with animals—where there can be no placebo effect.

Yet healing works. I have seen remarkable results obtained in a number of life-threatening circumstances. Most authentic healers actually do not know what they do; they know only that they have the "gift" of healing. They do not question *how* it is accomplished. Most healers are ordinary people who hold regular jobs and who do their healing treatments in their spare time.

The late Olga Worrall was one such person. "Auntie" Olga viewed her gift in a matter-of-fact fashion. In her later years, she conducted one healing session a week, in the basement of a neighborhood church. Each patient was given the necessary time, which was generally less than ten minutes. Olga presented the appearance of nothing more than a kindly grandmother, an ordinary person—yet her patients reported experiencing a feeling of great calm and contentment and a sensation of having received "something." Olga told me that the healing seemed to take something out of her; as she grew older, she had to limit her treatments because of her increasing fatigue. Something other than the simple placebo effect seemed to be occurring.

Since we know that the body uses electrical control systems to regulate many basic functions and that the flow of these electrical currents produces externally measurable magnetic fields, it does not require a great leap of faith to postulate that the healer's gift is an ability to use his or her own electrical control systems to produce external electromagnetic energy fields that interact with those of the patient. The interaction could be one that "restores" balance in the internal forces or that reinforces the electrical systems so that the body returns toward a normal condition.

In the past, we had only anecdotal evidence to support this concept. One "experiment" involved the healer placing his hands on the outside of a cloud chamber, a device originally used to detect the passage of high-energy atomic and subatomic particles. Strange things were said to occur within the chamber, but they were poorly characterized. This was further confused by a later experiment in which the healer, located many miles away, simply *thought* of the cloud chamber, and the strange events were said to recur. This experiment, while interesting and worthy of duplication, was uncontrolled, and the results have done little to clarify the issues.

Unfortunately, the intellectual bias against the phenomenon of the healer is so great that it has been extremely difficult, if not impossible, to get reputable scientists with the necessary equipment to become involved in valid studies. This condition is now changing. The healer phenomenon has become accepted by much of the public, and some forward-looking physicians are beginning to think that this technique, as well as the other techniques of energy medicine, should be scientifically studied.

At this most appropriate time, a young healer named Mietek Wirkus has come to the United States from Poland, where the situation for healers is much different. There, the healer is accepted as a valid medical therapist and is subject to licensing by the government. In order to be approved and licensed, a healer must demonstrate to a council of physicians that his or her treatments have actually worked on a specific number of patients. Wirkus and his wife arrived in the United States about two years ago.

Wirkus is convinced that his treatment involves the flow of energy from him to the patient, and that the energy involved is electromagnetic. This is particularly important because he, in contrast to Olga Worrall, does not touch the patient. Therefore, any energy involved must be of a type that is transmissible across space. Of the forces available, only electromagnetism qualifies.

I first met the Wirkuses early in 1988 and had the opportunity to ask Mietek a number of questions that I considered important from the energy-transfer point of view. In particular, I felt that if his ability involved a major level of control over his internal electrical control systems, then in addition to being able to "project" an external field, he should also be able to sense the disturbed electrical fields within the patients at the site of disease. In that sense, he should be able to make a diagnosis, not of a specific disease, but of the *site* of the disease.

In making this prediction, I drew on both theoretical science and my own experiences as a practicing physician and surgeon for many years. I have become convinced that some of us make a final diagnosis based not only on the physical examination and laboratory tests, but also on a "gut feeling" or intuition of some sort. It takes time to do this—time to talk with the patient in addition to conducting the physical examination.

The best example I can think of to show this in action involves the differential diagnosis between acute appendicitis and acute mesenteric adenitis. These two conditions have practically identical laboratory findings and physical signs on examination, yet the former must be treated surgically, while the latter will resolve itself without surgery. The modern physician will, more often than not, opt for the diagnosis of mesenteric adenitis and not operate, so that the stigma of doing an appendectomy when there is no appendicitis may be avoided. Of course, if the patient really has acute appendicitis, the appendix will rupture, and it can be treated (sometimes) with surgery and antibiotics. In contrast, I have seen older surgeons go into a patient's room, spend an hour, and emerge with a diagnosis. They were right far more often than they were wrong. I would suggest that the best surgeons thus make their diagnoses by some intuitive method that we do not yet understand.

When I spoke with Wirkus, he assured me that he could easily determine the presence of a disease and determine where it was located in each patient. In fact, even when he is first provided with a definite diagnosis, in treating any patient he first "scans" the total body, holding his hands a few centimeters away from the body surface, looking for other pathology. He was not sure exactly how he did this, but he felt that he sensed energy coming from the area of disease. He was also certain that when he treated a patient, some energy passed from him to the patient. He told me that it took much more energy from him to treat cancer or schizophrenia than to treat arthritis, skin conditions, insomnia, or neuroses.

I was able to watch Wirkus treat patients, and I also had the opportunity to have him work on me. The process appears simple: the fully clothed patient stands erect, and Wirkus moves about the patient, keeping his hands a few centimeters away from the body surface. He appears to be almost in a trance state; his eyes are open but appear unfocused, and he has a peculiar respiratory pattern of short, audible, regular breathing. The process takes less than ten minutes. When I was the patient, I shut my eyes so that I would not know where Wirkus's hands were. I wanted to determine whether I would experience any sensations during the procedure in any particular area of my body, without being influenced by my knowing where he was working. On several occasions I noted distinct feelings of warmth and tingling. Each time, on opening my eyes I found that the sensations were located near the position of Wirkus's hands. Immediately following the procedure, I felt his hands. They were distinctly cool and did not produce a feeling of warmth on my skin.

Wirkus had no knowledge of any of the diseases that I had. I have minor arthritis in my right hip joint, a mild spastic colitis, and rather severe glaucoma (with loss of almost 70 percent of the visual field in my right eye). On examining me, he diagnosed mild arthritis in the right hip, stated that I had some problems with my colon, and told me that the energy level on the back of my head was very low. While he found no problem with my right eye, the visual images from that eye go to the visual cortex at the back of the head. In the same session, he diagnosed several other individuals with similar precision.

Following my experience, we discussed the possible energetic aspects of his practice. He then described a singular event that reinforced the likelihood of such aspects. Before coming to the United States, Wirkus and his wife had given demonstrations of healing at various cities in Poland. One evening they were in a small city, using the stage of the local theater. The patient sat on a chair on the stage, with Wirkus standing next to him. Mrs. Wirkus was seated at a table, also on stage, about fifteen feet away. She had a microphone with which to explain to the audience what her husband was doing. The stage lights were located on the ceiling over the first row of seats, with the control box located in a cabinet at the side of the stage. The first three patients had simple depressive neuroses, which were easily treated, but the fourth case was a patient with cancer.

Because it was evening, the stage lights were on, but those in the rest of the theater were off. About fifteen minutes after treatment of the fourth case had begun, Mrs. Wirkus noted that the lights appeared to be slowly pulsing, and that this seemed to be producing waves of light over the front row of seats and the stage. The pulsing increased in intensity and frequency, and it was noted by the audience as well. The microphone then began to pulse at the same frequency, and Mrs. Wirkus pushed it away from her. The stir in the audience and his wife's movements caused Wirkus to stop the treatment, and the lights and sounds suddenly returned to normal. The theater electrician, who had been in the audience, came rushing up, asking, "What have you done to my lights?" This was the only time such an unusual event had occurred during a treatment, and the only time those particular stage lights had behaved in such a fashion in that theater.

Using a frequency generator and a speaker, I was able to have Wirkus and his wife determine the frequencies involved. They both felt that the oscillations had begun at about one per second (1 Hz), increasing in strength and frequency to about four or five per second (4–5 Hz), at which time Mrs. Wirkus pushed the microphone away and the session terminated.

It is, of course, impossible to determine exactly what happened, since all of this occurred several years ago and many miles away. However, we can speculate, using a knowledge of how stage lighting was done some years ago. The main requirement for stage lighting is that it be smoothly dimmed to full darkness and able to be "brought up" to full brilliance in the same fashion. In those days, this could not be done with the standard AC power system, and stage lights used a DC supply that could be slowly dimmed. If the theater had had such a system for stage lighting, it's possible that Wirkus gave off a slowly varying electromagnetic field that was strong enough to produce a "modulation" of the DC supply to the lights. From a physics and engineering point of view, this is the only tenable possibility, and this observation lends some support to the theory that the healer phenomenon directly involves electromagnetic energy.

Recently, there has been firmer substantiation of this theory through some experiments done in mainland China. The healer tech-

nique in China is called *Chi Gong* (or *Qi Gong*, depending on the
district in China that is involved). The method is more spectacular
than that used by Wirkus. The Chi Gong "master" stands several
feet away from the patient, makes certain classical, formalized
movements of his body and arms, and then points his outstretched
arm at the patient. The Chinese believe that the master acquires *chi*
energy by means of these movements, and that he then projects that
energy toward the patient, who is considered to be deficient in *chi*
or else has the two components of *chi* imbalanced. While to Western-
ers this may sound like nonsense, this therapy has, along with acupuncture, persisted for many thousands of years in China. Chi Gong
practitioners are also often called upon to treat animals, apparently
with considerable success.

　　Given the details of the treatment, the only defensible possibility is that there is a transfer of electromagnetic energy from the Chi
Gong master to the patient. In this case, what is the purpose of the
stylized movements that are considered so important to the therapy?
Dr. Jame Ma, professor of physics at the Chinese University of
Hong Kong, has made an interesting suggestion. He postulates that
these bodily movements are in the specific frequency range at which
the proton, the nucleus of the hydrogen atom that is so common in
water molecules of the body, will absorb energy from the Earth's
natural magnetic field by means of nuclear magnetic resonance
(NMR). (Because the concept of electronic resonance is so important
to the understanding of the relationship between living organisms
and electromagnetic fields, it will be discussed in detail later in this
book, in chapter 10.)

　　To explore the relationship between electromagnetism and the
Chi Gong phenomenon, doctors at the Huazhong Normal University
in Wuhan, China, used NMR in an attempt to determine whether Chi
Gong masters give off any electromagnetic radiation. They studied
the effect of Chi Gong "treatment" on the complex, bioactive, organic, phosphorus-containing chemical o-n-propyl-o-allylthiophos-
phoramide. This particular chemical was chosen because it produced
a well-characterized NMR spectrum in its normal state. However, if
exposed to a low-strength electromagnetic field, the chemical would
absorb the energy, and certain atomic "bonds" in its structure would
be altered. This change would result in a specific recordable change
in the NMR spectrum. The extent of the change in the NMR spectrum indicated the amount and the site of structural change in the

chemical molecule. This changed molecular structure would persist for several hours after exposure of the chemical to the electromagnetic field.

In the experiments, the chemical substance was enclosed in a sealed glass container, and Chi Gong masters were asked to "treat" it, holding their hands a certain distance away from the container. The NMR spectrum was recorded before the treatment and found to be normal. Following the treatment, the NMR spectrum changed significantly; the extent of the alteration in the NMR spectrum could be increased by repeating the Chi Gong exposure.

Similar studies are ongoing in the United States at this time; however, only preliminary results have been obtained. These studies are models of the type of rigorous, objective, scientific investigation that should be done in regard not only to the healer phenomenon, but to many of the other techniques of energy medicine as well.

The conclusion that may be reached from the Chinese studies is that the healer phenomenon has a basis in physical reality, and

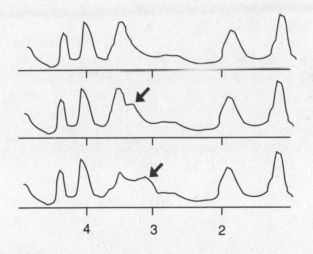

FIGURE 4-2. *NMR scans of the chemical.* Top, *natural-state NMR;* middle, *after first Chi Gong treatment;* bottom, *after second Chi Gong treatment. The resonance that had peaked at 2.5 before treatment broadened and showed a new peak (arrow) closer to 3 following the first treatment. The second treatment resulted in a decline in the original 2.5 peak, and the second peak increased in size and moved closer to 3. These NMR changes could be related to an increasing alteration in a specific part of the structure of the chemical molecule.*

that some form of electromagnetic energy is unquestionably involved. The results indicate that the healer gives off electromagnetic energy from his hands during the treatment process. It is essential that we determine the strength and the frequencies of the field produced by the healer.

An extremely useful device for this would be the SQUID magnetometer, the very sensitive magnetic field detector that made it possible to discover the magnetoencephalogram. While it is possible that the extreme sensitivity of the SQUID is not required, its accuracy and ability to measure both DC and time-varying magnetic fields makes it the best available. Unfortunately, the price of a good SQUID is about $75,000, and this does not include costs of normal upkeep. Since there are a number of such devices in use at various academic institutions around the country, it could be expected that in the interest of science, one of these devices would be made available for the few weeks necessary to complete this type of study. Unfortunately, this has not yet been done, despite a number of requests.

It also might be fruitful to determine those frequencies of the magnetic field to which healers are particularly sensitive. It may be possible to produce a "spectrum" of sensitivity that might indicate that different disease processes produce specific frequencies, and that these are sensed by the healer when he makes the diagnosis. If this should prove to be so, one could use this finding as a basis for designing diagnostic devices that would support, or even bypass, the healer for diagnosis. Wirkus may thus be cooperating in the ultimate demise of his technique—but, knowing him, I believe he would think it well worth it!

The minimal-energy techniques, which are based upon activation of the body's innate internal control systems, work. As an orthodox trained physician and board-certified orthopedic surgeon, I have absolutely no hesitancy in recommending that patients with life-threatening conditions follow this route after they have been given a careful and full disclosure of all the therapeutic possibilities open to them. In my opinion, the use of these techniques only as a "last resort"—after orthodox treatments have reduced or destroyed the patient's own resistance systems—is probably much less effective.

I try to follow one primary rule: make sure that the patient knows all there is to know about the illness and the therapeutic possibilities, both orthodox and unorthodox. Then, the patient must make his or her own decisions.

The only problem with the minimal-energy techniques is that at present we know so little about the mechanisms involved that we are unable to increase the efficiency of treatment or to shorten the time required for most such techniques to be effective. A patient who is determined to follow minimal-energy techniques to deal with a life-threatening illness must know that, while this course of treatment may be successful, it requires the expenditure of large amounts of time. Despite being convinced of the efficacy of these techniques, many busy people simply cannot follow such a course. Recognition of this problem has led to an increased interest in techniques that involve the administration of artificially generated electrical or magnetic forces, so that the internal control systems can be influenced immediately. These techniques will be our next topic.

CHAPTER FIVE

BUILDING UP THE BODY'S ELECTRICAL SYSTEM: ENERGY-REINFORCEMENT TECHNIQUES

The cell is a machine driven by energy. It can thus be approached by studying matter, or by studying energy.

ALBERT SZENT-GYORGYI, *Biolectronics*

We have seen how the body's internal energetic systems may be accessed by the conscious mind through the use of several techniques that do not involve the addition of any external energy into the body. The body's internal energetic control systems are subtle, and they operate with minute amounts of electromagnetic energy. A little goes a long way, and very often, more is *not* better. In the field of energy medicine, we would do well to follow the old aphorism to assist rather than replace nature. Much commitment and enthusiasm will obviously be required for a successful transition from technological medicine to energy medicine. But our enthusiasm must be tempered with prudence and based upon knowledge of how such techniques work and what their potential side effects may be. It is evident from the preceding chapters that electromagnetic forces lie at the most basic levels of living things—and that this is an area we should enter with caution.

This chapter will deal with those techniques that appear to work by adding to, or reinforcing, the existing internal energetic systems through the application of small amounts of external energy. While these techniques introduce external energy and are, therefore, potentially more dangerous than the minimal-energy techniques, they are less dangerous than techniques that employ the same forces at magnitudes that overwhelm the internal systems. Again, I will limit my discussion to those techniques with which I have had personal experience and that have the theoretical capability to act via electromagnetic forces.

HOMEOPATHY AND THE LAW OF SIMILARS

If body energies are subtle, then the subtlest active treatment involving energy has to be homeopathy. This concept of medical therapy actually began in the early 1500s with Paracelsus' "Law of Similars," which postulated that one could cure a symptom by giving very minute quantities of a medication that would produce the same symptom in larger doses. At that time this "symptom-oriented" approach to medicine was perfectly acceptable, since the causes for diseases were unknown. We can only speculate that Paracelsus based his idea on his intuition and on subsequent experimentation. It remained for Dr. Samuel Hahnemann to organize this concept into a coherent system of medical practice, more than 250 years later. Hahnemann published his major work, *Organon on the Rational Basis of Healing*, in 1810, at a time when chemistry was in its ascendancy and medical prescriptions were lengthy and complex and often included large doses of toxic drugs.

Hahnemann's prescriptions, consisting of extremely small doses of single preparations, greatly simplified medical practice and attracted a large following. Homeopathy was quite popular during the nineteenth century, when it served as a harmless alternative to the purging, bloodletting, and toxic drugs popular among the physicians of that time.

Hahnemann organized homeopathy by developing the method of "provings," in which he investigated the physiological effects on normal human subjects of large doses of many different natural materials. In this way, he built up his "Materia Medica"—a compendium of symptoms and the minute doses of remedies that cured them.

In modern homeopathic practice, a lengthy interview is an essential element. The physician tries to evaluate the totality of the patient and his or her reaction to the agent that is presumed to be causing the symptoms. Even though many homeopathic remedies have the same purported physiological effects, the homeopathic physician tries to select the precise one for each patient's psychological type. This is in accord with Hippocrates' concept that the symptoms of each disease are a combination of the causative agent *and* the patient's reaction to it. Such a "personalized" type of therapy must contain a large element of placebo effect.

Modern scientific medicine, on the other hand, is disease oriented. Therapeutic agents are selected solely on the basis of their antagonistic action against the disease-producing agent. This concept originated as science began to identify those outside agents, such as bacteria, that were causes for disease. In the case of infectious diseases, this has been remarkably successful, starting with aseptic surgery and then moving to public-health measures concerning contagious diseases. More recently, antibiotic therapy for established infections changed medicine from an "art" to a "science." However, with the advent of internally produced, degenerative diseases, the application of this simple causal relationship totally neglects Hippocrates' ideas.

Homeopathy is the other side of the coin. Being symptom oriented, it concentrates its attention primarily on the body's reaction and neglects the causative agent.

Technological medicine and homeopathy accuse each other of lacking an appropriate scientific basis. Since technological medicine has been so successful in the past, its pronouncements of scientific authority carry more weight today. It considers the clinical evidence for the effectiveness of homeopathy to be anecdotal and solely the result of the placebo effect. This criticism is reinforced by Hahnemann's technique of multiply diluting his preparations, which was so drastic as to appear to be absurd. He arrived at solutions in which *no trace* of the original medication could possibly remain. To complicate matters, Hahnemann also subjected each series of dilutions to "succussion," a methodical, mechanical agitation produced by banging the container of medication on a tabletop a certain number of times at each dilution. In Hahnemann's terms, the mechanical agitation "potentized" the material, or made it more effective.

Homeopathic preparations are thus characterized by the extent of the final dilution and the amount of mechanical agitation, with greater dilutions and more mechanical agitations resulting in a more "potent" medication. Chemical scientists believe that not only are the dilutions so great that no trace of the active agent remains, but also that the mechanical "succussion" has absolutely no effect on any chemical species.

Finally, the homeopathic remedies were, in chemical terminology, poorly characterized. The actual preparation of most such remedies involved chopping up an entire plant, roots and all, soaking it in alcohol for a number of weeks, and then subjecting the alcohol extract—or "mother liquor"—to the dilutions and succussions. Obviously, even in low dilutions the solution contained literally thousands of different chemicals. Therefore, in the rush to base medicine entirely on established scientific principles, homeopathy was decreed to be just so much hogwash.

The only problem with that judgment is that homeopathy works well enough clinically that, despite savage attacks by scientific medicine, it has survived. Its survival can be attributed in part to the safety of homeopathy compared with the bleedings, purgings, and toxic medications of "real" medicine in the 1800s. However, this cannot have been the whole story. The practice of medicine depends to some extent upon the pressures of the marketplace, and homeopathy must have been sufficiently useful to retain a market position until today. While those physicians who practiced it insisted that it worked, the scientific physicians decried its use as preventing patients from obtaining "proper" treatment. Yet anecdotal reports abound, attesting to the fact that orthodox physicians often referred not only their "difficult" cases to the homeopath, but members of their own families as well. Despite all the theoretical objections in the world, homeopathy often appears to work.

Over the past decade, there has been a marked upsurge of interest in homeopathy, with increasing numbers of scientifically trained physicians beginning to learn the techniques and actually use them in their practice. As a result of this increased interest, there have been a number of attempts to place homeopathy on some sort of scientific footing.

A truly scientific evaluation of the clinical effectiveness of homeopathy was not done until 1986, when Dr. David Reilly and his coworkers at the Glasgow Homeopathic Hospital reported in the prestigious British medical journal *Lancet* on an impeccable, double-blind study of the efficacy of a homeopathic preparation for hay fever compared with an ineffective placebo. They reported that "the homeopathically treated patients showed a significant reduction in patient- and doctor-assessed symptom scores." In addition, they found that "the significance of this response was increased when results were corrected for actual pollen counts, and the response was associated with a halving of the need for antihistamines." Obviously, the homeopathic preparation had a definite physiological effect that cannot be completely explained by the placebo effect. This occurred despite the fact that the medication was diluted to a point at which, theoretically, none of the original medication remained.

The publication of this paper elicited a storm of letters to the editor, which were interesting for what they revealed about the mind-set of the modern "scientific" physician. Despite the study's having been an exceptionally well-planned and well-executed scientific evaluation, one writer called it "the first randomized, double-blind trial of one placebo against another," and went on to state that "the homeopathic potency used contained not one molecule of the original extract." The objections raised by the critical letters came down to one point: homeopathy simply cannot work. If the results of a scientifically valid study are at variance with established theory, the study must be incorrect. The idea that one should consider revising the theory was not mentioned, except by a few correspondents.

The Benveniste Affair: A Research Scandal

Most recently, a study published in the British journal *Nature* has elicited an even greater storm of controversy, not only over the data presented but over their political implications as well. Dr. Jacques Benveniste and his colleagues at the University of Paris reported on a study of a type I had long advocated.

The problem of dealing with any possible mechanism of action using dilute solutions is compounded by the variable and complex chemical nature of homeopathic preparations. I have therefore urged that *single* chemicals of a known biological action be subjected to the same dilutions and mechanical agitations as homeopathic preparations, with tests for their biological actions being made after each dilution-succussion. Dr. Benveniste used antisera produced against immunoglobulin E (IgE), the human hypersensitivity antibody. In the test tube, the reaction between the antiserum and IgE produces a specific response that is revealed by staining certain human white blood cells with specific dyes. While other, simpler systems could have been used, I believe that Dr. Benveniste used this system because it is the basic biochemical process in the allergic reactions that were the subject of Reilly's double-blind experiment.

Benveniste found that as dilutions increased beyond the point at which there were theoretically no molecules of antiserum present, successive peaks of cellular effect were observed. In other words, the cellular effects appeared at some dilutions, disappeared at higher dilutions, and then reappeared at still higher dilutions that were well beyond the theoretical limit at which the last molecule of original chemical was still present. He concluded that either some of the anti-IgE antibodies were still present even at these impossible dilutions, or that the "essence" of anti-IgE had been impressed upon the water molecules in some fashion.

When he submitted his paper to Dr. John Maddox, the editor of *Nature*, it was decided that Dr. Benveniste had to obtain confirmation of the effect from several other laboratories of his choosing. This process consumed two years. When the other laboratories reported the same findings, *Nature* published the paper, along with an editorial comment that there was no present physical basis for the observations and that "there are good and particular reasons why prudent people should, for the time being, suspend judgment." On the face of it, this is a reasoned scientific statement; however, on reading the entire editorial one finds that the "good and particular reasons" are the present laws of physics, such as the law of mass action. One does not find any indication that these "laws" might require some adjustments when dealing with such extreme dilutions of active substances. In short, what was being protected was simply dogma.

The system of scientific checks and balances following the publication of a controversial paper consists of other scientists attempting to verify or disprove the reported findings. This cannot be done in this case, because Maddox's editorial has "tainted" the subject. No established scientific researcher would risk his or her reputation by attempting to reopen the controversy.

Concurrent with the publication of the paper, Benveniste was notified by Maddox that an investigatory team was going to visit his laboratory. The team would be composed of Maddox (a physicist), Dr. Walter W. Stewart, a self-appointed "scientific fraud investigator" from the U.S. National Institutes of Health, and James ("The Amazing") Randi, a professional magician recently turned scientific fraud detector. Maddox's position had, of course, been spelled out in advance in his editorial; Stewart had no specific qualifications in immunology, which is how the Benveniste study would have been classified; and Randi lacks any scientific credentials. It is obvious that Maddox suspected that Benveniste's study was fraudulent. A reputable scientific journal either rejects a paper or notifies a scientist's home institution of its suspicions. In neither case is a paper published until the suspicions are resolved.

According to the investigative team, they found poorly controlled conditions in Benveniste's lab, and when the experiments were performed for them, the results produced were no better than chance. In his letter of rebuttal, Benveniste wrote that the team had behaved in a totally unscientific fashion while at his laboratory. Positive findings were disregarded, his personnel were prevented from doing their work adequately, and the atmosphere was that of a witch hunt. The story has been widely reported in the world press, generally with a mocking attitude toward the data presented ("The Water That Lost Its Memory," *Time* magazine, 8 August 1988).

Dr. Maddox has adopted the stance of defending the laws of physics, but what he really appears to be doing is defending the biomedical status quo against a "heretical" concept. Somewhere in all of this, the spirit of true scientific inquiry has been lost. Is it best to defend the status quo and reject new scientific data simply because these data may pose a challenge? Or should we seek new data and revise our "laws" accordingly? The history of science is one of progress made after the latter path has been followed, no matter how difficult this may have been.

A NEW ANGLE ON THE SCIENTIFIC BASIS FOR HOMEOPATHY

The key problem in homeopathy is the dilution of the active solution far beyond the point at which even a single molecule of the original substance remains. Many theoretical attempts have been made to explain this, usually on the basis that the solvent molecules—in this case, water—undergo changes in configuration and somehow become "imprinted" with a specific pattern imparted by the homeopathic material. The difficulty with this approach is that the water molecules would have to be able to assume as many different conformational changes as there are specific effects of homeopathic remedies—not a very likely occurrence.

The importance of succussion suggests that the molecules of the active material may be the ones that are changed, by disruption of bonds attaching the active groups to the parent molecule. The active groups may consist of a few atoms in a specific configuration that act by binding to the cell membranes. Such small groups of atoms are usually surrounded by a shell of "structured" water molecules, which have properties that are different from those of the rest of the solvent. This hydration shell of structured water may itself have a configuration that reflects that of the active group. In this fashion, the solvent water may theoretically be considered to have been imprinted.

Twelve days after Benveniste's paper appeared, a paper published by Jacob Israelachvili and Patricia McGuiggan of the University of California at Santa Barbara reported on some complex experiments dealing with the forces between surfaces in liquid solutions. The authors were able to measure oscillatory forces of attraction and repulsion related to a number of factors (such as the concentration of the solution) at very small distances from solid surfaces. They made the point that these types of forces may have fundamental implications for the biology of cell membranes, and that much remains to be learned about them.

In short, we really do not know very much at all about energetic actions over extremely short distances in very dilute solutions. The law of mass action may hold very well for the usual case, but it works poorly, if at all, in a very dilute solution that has been subjected to mechanical agitation.

Obviously, when we study the actions of highly dilute solutions and the forces acting on solutions at the level of the cell membrane, we are involved in an area in which our knowledge is far from complete. The structure of most biologically active molecules is exceedingly complex, and relatively few investigators have been determined enough to plot structural maps of their atomic relationships. In solution, these actions are even more complex.

In contrast to our usual idea of the molecule's simply "floating about" in the solvent, the entire molecule is in a state of constant agitation known as Brownian movement, which results from the thermal agitation of the smaller, solvent molecules. Further, a complex biological molecule is itself constantly changing. Some molecules fold and unfold about certain chemical bond angles, and active groups are rotating about their bonds literally millions of times per second. To this we must add the constant motions of electron clouds about each atom of the entire molecule. The beaker of solution that seems to be sitting placidly on the table is actually teeming with activity.

Modern technological medicine, in the main, looks upon therapeutic drugs as agents that attack other "agents" that have produced disease. Since the body itself plays no major part in this scenario other than being the place where the action occurs, one has to administer enough of a drug for it to act effectively. In chemical parlance, one adds the drug "to excess," so that the reaction with the disease-producing agent proceeds "to completion." This idea is well and good for the situation in the test tube, but clinical medicine is not conducted in test tubes.

If we accept the point of view that the body can heal itself through its own energetic systems, things seem much different. The cells of the body respond to energy in the form of electron transfer. The electrons are carried by chemical structures that fit into matching chemical structures on the cell's surface. These chemical structures may be small parts of a much larger molecule, but the active group that carries and transfers the electron may consist of only a few atoms in a particular geometric configuration. This concept has been partially explored and confirmed, particularly in the case of enzymes, immune bodies, and certain toxins.

If we look at homeopathy from the perspective of bioelectronics, a different picture of its mechanism of action emerges. How many molecules of a substance are required for a biological effect? If we use the standard pharmacological concepts, we need a respectable number of molecules. However, if we're dealing with biological chemical-sensing systems, fewer molecules are necessary. For instance, pheromones are complex chemical sex attractants that are released into the air, particularly by insects. Obviously, the total number of molecules is small, and that number is rapidly diluted to an enormous extent by the air and by wind currents. Yet they work very well; male moths respond to such unseen cues from miles away. It has been found that even one molecule of the appropriate pheromone will trigger the appropriate neural circuits, causing a male moth to fly toward the female. When you consider that the essence of this action is only a single active group and the transfer of a single electron, it becomes even more remarkable.

Again, the biological factor enters into the equation. Living organisms have developed reactive systems of exquisite sensitivity and selectivity. While this illustration is drawn from our present knowledge of the olfactory system in insects, similar sensitivity and selectivity can be demonstrated in the immune system of "higher" animals. The homeopathic preparation need only be capable of transferring a single electron to the appropriate receptor group on the cell surface in order to call forth a major biological response.

A final consideration in our search for a rationale behind the action of homeopathic preparations is the body's own reaction systems. I have been struck by the fact that in most homeopathic treatments the first reaction is a worsening of symptoms, with symptomatic relief coming later. To the homeopathic physician this initial stage is encouraging; it's a sign that he is on the right track. It is significant that the experimental group of patients in Reilly's study, cited above, demonstrated this effect. Since this was a careful, double-blind study, the reaction cannot be attributed to a placebo effect.

This phenomenon has received little study, yet it could contain a major key to the mechanisms involved. What functional systems are involved in it? Very likely the immune system is a major contributor, and its actions can be fairly easily monitored. It would be significant if in the course of another, similar double-blind study, indices of immune-system reaction were found to increase during the period of treatment.

In actuality, any studies of the type performed by Reilly can only support the clinical validity of homeopathy; they cannot reveal the mechanisms of action. Studies of the type performed by Benveniste can come closer to the mechanisms involved by showing that *something* active still persists even in impossible dilutions. However, Benveniste's choice of a biological assay (a reaction of living cells) to demonstrate this is technically difficult. There are other methods of assay, such as the electronic resonances, that could yield more objective results. The primary problem is to overcome the innate prejudice against homeopathy so that adequate studies can be done.

HOMEOPATHY AND TRADITIONAL PHARMACOLOGY

On a more mundane plane, homeopathy has something else to offer to traditional pharmacology. Many of the homeopathic preparations in common use today contain active materials that appear to be much more effective than the synthetic drugs used for the same conditions. In my experience, the homeopathic ointment Arnica (prepared from the leopard's bane plant) has remarkable local pain-relieving, antiinflammatory, and fibrinolytic (clot-dissolving) actions that are far more effective than those of any standard pharmacological agent. Arnica ointment, however, is not a highly diluted, succussed preparation, and it does contain measurable amounts of a large number of substances, some of which have truly astonishing clinical properties. Its local effects are due to some direct chemical action. Arnica is also available as a standard diluted and succussed preparation that is used systemically and is purported to be equally effective, but I have had no experience with it in this form.

In many years of orthopedic practice, I have treated hundreds of sprained ankles. This condition involves a tearing of the soft tissues, such as the joint capsule and ligaments, resulting in pain, swelling, and hemorrhage (the purple "bruise") into the tissues. I have seen fads in treatment for this condition come and go: injecting the area with novocaine and making the patient walk, putting the leg in a cast, wrapping the ankle with an elastic bandage, and giving the patient cortisone. The least harmful of these treatments, putting the leg in a cast and gradually permitting the patient to walk, results in a healing time of at least three weeks; it takes even longer for the discoloration of the blood clot to disappear. The most harmful treatment, injection of novocaine (which is no longer advocated), results in recurrent pain and swelling, and, often, a complete tearing of the ligaments when the patient walks on the ankle. This treatment has frequently resulted in the ankle's becoming unstable and requiring corrective surgery.

Provided that the ligament is not completely torn, I have found that simply rubbing on Arnica ointment within a few hours of the injury results in practically instantaneous and complete relief of pain, a rapid and complete resolution of swelling, and a rapid (one to two days) disappearance of the blood clot. The average patient treated in this way is totally functional within five to six days. The most remarkable aspect of all this is the efficiency with which the blood clot is mobilized and dispersed. I know of no other agent, FDA approved or otherwise, that can match the efficiency of Arnica in this respect.

As this is being written, the newest wonder drug is tissue plasminogen activator, or TPA, a chemical agent made using recombinant-DNA techniques. TPA dissolves blood clots in the coronary arteries to the heart, the major cause of heart attacks. While there is no question that it has saved lives, the drug is extremely expensive and may also have serious side effects. Since Arnica has been around for centuries, and its effects have been observed by many thousands of physicians, one can reasonably ask why *its* clot-dissolving effects have not been investigated in a responsible scientific manner. Local application of Arnica ointment works so well that it would seem desirable to evaluate its systemic preparation for this purpose.

There is another observation on the action of Arnica that may be of great importance to energy medicine and to our idea of the DC electrical system as being responsible for the conscious perception of pain. The immediate relief of pain by Arnica depends upon the ointment's being present on the skin at all times. If it is washed off, the pain returns immediately, but it can be relieved just as rapidly by simply applying fresh ointment. Even though the source of the pain is in the deeper soft-tissue structures, such as the ligaments, the rapidity of the pain relief would seem to require that the ointment's action is upon some structures in the skin proper.

Could these structures be the neuroepidermal junctions that we previously proposed as the anatomical structures in the acupuncture points? If this is correct, Arnica may act by blocking the action of these junctions, as acupuncture needles inserted into the same points do. Admittedly, this is still speculative, but it does provide a theory upon which to begin experimentation.

It makes sense that what we perceive as "pain" is actually the signal that an injury has occurred. Since the skin is the junction between us and our environment, whatever pain "receptor" system is used should be located in the skin. Since local injury also produces reactions of swelling and hematoma formation in the tissues immediately below the skin, one may also postulate the existence of some *local* DC system associated with "pain receptors," which might act to mitigate these reactions. Certainly, if one were designing an organism, such a local system would be desirable from the point of view of survivability of the injured organism.

Local pain may also be partially relieved by applying pulsed electrical current to the area. There has been some clinical application of this using transcutaneous electrical nerve stimulators (TENS) units; these will be reviewed in the next chapter. I have found that both Arnica and TENS units are more effective for relieving pain in soft tissues than in bone. The pain of arthritis, for example, does not respond well to either agent. At this time I cannot provide any explanation for this phenomenon. At any rate, further investigation of the complex relationships among pain, acupuncture, Arnica, and local pulsed electromagnetic stimulation would seem to be a good idea.

I have found another homeopathic preparation to be equally useful in the local treatment of minor to moderate burns. The ointment Urtica Urens is derived from extracts of the common stinging nettle plant. Local first- or second-degree burns of the skin result in pain, swelling, inflammation, and blister formation. Application of the Urtica Urens ointment before blisters have formed results in immediate pain relief, the disappearance of swelling and inflammation within one to two hours, and prevention of blister formation. Like Arnica, Urtica Urens ointment must be kept on the injured area continuously and must be replaced if washed off. This treatment is much more effective than are treatments with topical cortisone or any other such agent. (Obviously, third-degree burns or extensive first- or second-degree burns require the immediate attention of a physician).

ACUPUNCTURE: TRADITIONAL VERSUS ELECTRICAL

Acupuncture, like homeopathy, is a symptom-oriented therapy. This is perfectly understandable given that both were developed many years prior to the identification of actual disease agents.

Traditionally, the useful effects of acupuncture are attributed to the restoration of balance of the twin forces of yang and yin within the body. This mechanism of action is a philosophical concept that was developed long before we had scientific knowledge of how the body actually worked. Acupuncture therapy is totally empirical, with the system of meridians and points worked out over many years through trial and error. So while there may be truth and utility in the therapy, it may be encumbered by many nonessentials and anachronisms of dubious value. More recently, acupuncture has come to be viewed as a discipline of energy medicine, operating via electrical forces. Attempts have been made to integrate this view with the traditional philosophical concepts of this technique. Some confusion and a number of conflicting claims have resulted.

Earlier, I described how my colleagues and I became interested in this ancient technique and summarized the results of our experimentation. While we were engaged in this project, word began to come out of China that the technique could be simplified and made more effective by the application of *pulsing* electrical current to the acupuncture points. This did not seem to make much sense, in light of what we were finding. When a colleague appeared one day with one of the Chinese electrical acupuncture devices, I eagerly tested it. Perhaps the Chinese knew something I didn't!

I was horrified to find that the levels of current and voltage were very high—so high, in fact, that they stimulated contractions of the local muscle groups. With such large electrical forces being administered directly to the acupuncture point, electrolysis and cellular damage had to occur. I decided that the Chinese had fallen into the same trap that many others were about to fall into: if the system seems to work electrically, let's make it work faster and better by applying substantial amounts of external electrical current to it. In the absence of full knowledge of the operating parameters of the internal system, this is a dangerous thing, and it is even more dangerous when the basic physical principles of electricity are not applied.

The old technique of acupuncture is totally safe from this point of view. The insertion of a metallic needle into any part of the body will produce a very small electrical current, because the needle insertion produces a local current of injury and the metal of the needle reacts with the ionic solution of the body. The traditional "twirling" of the needle may produce a pulsing current at a very low frequency. Applying heat to the end of the needle may slightly increase the reaction between the metal and the body fluids. This may involve a little more trouble and work than the newer method, but it's a lot safer.

The theory that we derived from our experiments was simple, in accord with physical principles, and supportive of classifying acupuncture as one of the reinforcement techniques of energy medicine. The acupuncture points served as "booster" electrical amplifiers for the very small DC electrical currents flowing along the meridian. The actual points could be structures identical or akin to the DC-generating neuroepidermal junctions we had produced in the rat. The insertion of an acupuncture needle in close proximity to such a point would interfere with the electrical workings of that point.

This idea appears theoretically adequate to explain the ability of acupuncture to block the transmission of pain sensation. All that is required is a blockage of the incoming signal that indicates damage and that is perceived by the conscious mind as pain. However, what we found was only the first tantalizing glimpse of this system.

In order to put this concept on a firm foundation, we proposed to measure the passage of a signal along the meridian and to then attempt to block the signal transmission with the traditional needle insertion. If successful, such a study would go far to establish the validity of this technique. The experiments would provide a firm basis for further study of this intriguing system.

While I was certain that our data had proved that the acupuncture system does exist and has the capability of transmitting information, the proof that it was clinically effective was much more difficult to obtain. Despite much work by other NIH grantees and other groups, the problem of distinguishing the effects of acupuncture from a placebo effect remained dominant. Almost all of the

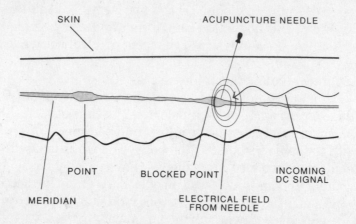

FIGURE 5-1. *A theory of how traditional acupuncture works, as derived from our studies. Since the system would use the minute levels of voltage and current that could be generated by the body, the small electrical field developed by the metallic needle is sufficient to prevent the treated point from amplifying the incoming DC signal and passing it on to the next point. While the points have not yet been anatomically identified, the evidence for their presence is excellent. No effective study has been done to locate the structures associated with the points identified by electrical parameters.*

studies were double-blind and were directed toward blocking or relieving pain. There are any number of studies in the literature showing that the injection of distilled water is often as effective in relieving pain as an injection of morphine, provided that pain relief is what the patient has expected. Since acupuncture involves the insertion of needles and the subjects expect pain relief, this could well have been what happened. We still have no way of objectively determining the presence or extent of pain, and so the placebo explanation has prevailed. Acupuncture, in the view of the medical establishment of the 1970s, was completely explained by the placebo effect. Despite a lack of objective scientific evidence, the clinical use of acupuncture is even greater today than it was at that time.

Our scientific approach would have enabled us to evaluate the other effects ascribed to acupuncture treatment, one of which is a measure of control over growth processes. While anecdotal reports commonly indicate useful effects upon tumor growth and remarkable healing of such degenerative conditions as arthritis, these reports are subject to a number of uncertainties. However, it is possible to theorize a connection between acupuncture and growth control, based on the concept of the closed-loop, negative-feedback DC system of data transmission. For example, if the output healing system is called into action by signals entering the cerebral DC control area, then appropriate stimulation of the acupuncture system could produce the same effect. At this time, we have absolutely no solid scientific proof for this theory or for its actual clinical utility.

While orthodox, scientific medicine has turned its back on acupuncture, veterinary medicine is becoming a vital and most interesting area for further study of this entire system. Obviously, veterinary medicine offers one great advantage over human medicine: the placebo effect simply cannot play a role in any effect that is observed and documented in animals.

Over the past fifteen years veterinarians have become increasingly interested in acupuncture as a pain-relieving and therapeutic technique. Results have been so good that an International Veterinary Acupuncture Society has been formed, and in 1987, the American Veterinary Medical Association (the counterpart of the AMA) came very close to officially recognizing acupuncture as a valid veterinary medical procedure. At present, acupuncture is being widely used in veterinary practice with great success, mainly as a way to produce general surgical anesthesia and analgesia for chronic painful conditions.

Surprisingly, acupuncture has been found to be very effective in the treatment of arthritic joints in animals. It has been reported to be clinically effective in increasing activity and relieving symptoms in 84 percent of animals that had degenerative joint disease and that were unresponsive to standard treatment. X-rays taken before and after treatment of the involved joints strongly suggest that regeneration of the damaged joint cartilage has occurred. This, however, has not yet been firmly established by autopsy. Similar results were reported in degenerative disk disease of the spine in animals. If autopsy or biopsy studies were done on these joints and disks, and actual healing or regeneration was demonstrated, we would have proof of the ability of acupuncture to favorably influence the output side of the control-system loop. The implications for human medicine are vast.

Finally, veterinary acupuncture treatment has been shown to be remarkably effective in idiopathic epilepsy (epileptic convulsions occurring without any evidence for physical damage to the brain), a condition that had been resistant to other therapies. This confirms clinical reports of the same effect in humans and, with regard to the total DC control system, is a very important observation. In both animals and humans, epilepsy is the spontaneous, synchronous discharge of a large group of adjacent nerve cells in the brain. This synchronization is the result of negative polarity shifts in the DC potentials in the area, which increase the excitability of the nerve cells. It is possible that the well-known "aura"—the premonitory feeling that a convulsion is about to take place—is the conscious perception of this DC change. The ability of acupuncture to effectively prevent epileptic convulsions in both animals and humans may well be due to changes induced in the cerebral DC system that result from appropriate changes in the incoming DC acupuncture system.

The problem facing veterinary acupuncturists is a lack of funds to pursue the necessary basic research that could actually prove what is happening. Since in clinical veterinary use the effects obtained by acupuncture cannot be due to any placebo effect, in addition to our having proved that the acupuncture system actually exists we also have totally objective evidence for important clinical efficacy obtained through the observations of veterinary physicians.

As a result, there has been a substantial increase in the number of practitioners of acupuncture. Since the technique is officially nonexistent, there are few regulations concerning the training and qualifications of these people. Traditional acupuncture is complex and difficult; it is not something that can be learned by anyone—including a medical doctor—in a three-day course.

Many of the courses offered today stress the idea of using fewer points and "treating" them with one of the many electronic devices now available. I believe that this is a mistake. While my data indicated that acupuncture is based on the flow of DC electrical currents, these currents were minute in strength, and I had barely begun to evaluate the complexities of this system. My findings are not adequate to justify the application of externally generated electrical currents to acupuncture points, and I do not believe that the Chinese have any more substantive data.

While there are any number of electronic devices sold that are purported to be able to seek out the exact position of acupuncture points and to treat them with exactly the right amount of electricity, these also are not subject to any quality control. A totally objective evaluation of these devices has yet to be done, but such an evaluation cannot be done until we have some additional knowledge of the electrical characteristics of the acupuncture system itself.

In short, the present situation is a mess. There are some physicians practicing acupuncture who are well trained in the traditional system. There are other practitioners who are not M.D.'s but who are equally well trained. Unfortunately, there are also many practicing acupuncturists of both types who are poorly trained. Lacking a basis of knowledge, they implicitly trust their electronic gadgets to do the treatment for them.

The best advice I can give to any prospective patient is in general to avoid those practitioners who rely solely on various electronic devices for diagnosis and treatment. If you wish to explore the possibility of acupuncture as a treatment, first be certain of your diagnosis and know the possible orthodox therapies available. Then, if you still desire to follow this route, seek out a traditional acupuncturist.

OTHER ENERGY-REINFORCEMENT TECHNIQUES: VITAMINS, TRACE ELEMENTS, AND PHYSICAL MANIPULATION

There are a number of additional techniques that might possibly be placed in the energy-reinforcement category. These include the use of dietary supplements, such as vitamins or trace elements, and the manipulative techniques of osteopathy and chiropractic. Of the vitamins, ascorbic acid (vitamin C) is the most prominent in the lay press. It certainly has a prominent proponent in Linus Pauling, two-time Nobel Prize winner. While this vitamin could theoretically play a role in the energetic and immune systems of the body, firm evidence for this is still lacking. The clinical trials of vitamin C on cancer patients are clouded with controversy. In my opinion, a valid trial has not yet been done. However, this area shows promise; in my practice, I have used moderately high doses of vitamin C on patients with chronic-fatigue syndrome and have observed useful results.

The trace-element hypothesis is interesting, but there is little firm data on it. It is possible that some of the metallic trace elements, such as cobalt, may play a role in the semiconducting structures in the body, but this also remains to be proved.

Manipulation therapy has enjoyed a long and clinically useful life, and many patients are convinced of its efficacy. It has been my observation as an orthopedic surgeon that some painful conditions of the complex joints may be relieved by appropriate manipulation. Once I was certain that such patients had no conditions for which manipulation therapy would be dangerous, I often referred them to an osteopathic colleague. Note that this treatment is simply mechanical and is not related to the energy systems of the body. However, there are many osteopaths and chiropractors who postulate that their manipulations somehow "balance" these energy systems. Unfortunately, objective evidence for this idea is lacking.

In the group of therapies that supplement the body's own energy systems with small amounts of external energy, there are several useful techniques with which some remarkable clinical effects have been demonstrated. These techniques appear to work by enhancing the intrinsic energy-control systems of the body, and, as such, they have some basis in scientific fact. But further exploration is required: the mechanisms of action of these techniques must be better determined, their actual clinical utility fully evaluated, and ways sought to make them more efficient and more reliable for larger numbers of patients.

With the growing popularity of such therapies, the danger of intrusion by fraudulent practitioners is great. The best defense against this is the establishment of a scientific foundation for the actions of all practitioners so that these treatment techniques may be accepted as valid and clinically useful.

Such techniques appear to be safe as long as they are employed in traditional fashions. There does exist the danger that they could be applied to conditions for which they are of no value, thereby delaying appropriate treatment, or that "modern enhancements" (such as electroacupuncture) could be added to them that might either defeat the original purpose of the therapies or render them dangerous. Modern clinical medicine seems to be searching for quicker, easier cures. When electromagnetism is involved, this can lead to overenthusiastic applications of excessive amounts of external energy, with potentially harmful results.

CHAPTER SIX

ADDING TO THE BODY'S ELECTRICAL SYSTEM: HIGH-ENERGY TRANSFER TECHNIQUES

The authorities of the London Cancer Hospital will be unfaithful to their honourable trust should they decline to test to the fullest extent the curative effects of frictional electricity in some of the most hopeless variety of diseases to which humanity is exposed.

A. ALLISON, M.D., *Lancet* (10 January 1880)

A few years following the Galvani–Volta controversy in the 1790s, Galvani's rather odd nephew, Aldini, treated a man for a psychiatric condition by using multiple applications of direct current to the patient's head. This is the first recorded example of electrotherapy with a possibly effective "dose" of electricity. Aldini claimed that the treatment was successful and that the patient was able to return to his normal activities. Given the state of the art at the time, the voltage and current used must have been small. Aldini could hardly qualify as an unbiased investigator, because he had also been involved with such activities as "revivifying" the dead through the application of DC current to fresh corpses in sufficient amounts to stimulate muscular movement.

At that time, electricity entered what could be called a therapeutic vacuum. The treatments available to physicians were mainly ineffective and were often toxic, with recoveries due only to the resiliency of the human body. Electricity at that time was a mysterious, little-understood force that was identified as the "vital spirit" of life itself. Obviously, it could therefore be considered capable of curing any ill.

In 1812, a surgeon by the name of Birch, working at St. Thomas' Hospital in London, reported that he had successfully treated a patient with a nonunion of a fracture of the tibia by using "shocks of electric fluid." By 1841, this bone-growth-stimulating technique had progressed sufficiently to be considered the treatment of choice for such conditions. Other disease states were not neglected during this period. In 1860, Dr. Arthur Garrett, a fellow of the Massachusetts Medical Society, published a textbook entitled *Electrophysiology and Electrotherapeutics*, with instructions for treating a range of clinical conditions using electricity. By the 1880s, many medical schools in both the United States and England had professors of "electrotherapeutics" on their faculties.

In less than a hundred years, electrotherapy had become a respectable part of the average physician's armamentarium. While the technique was not universally accepted by the medical profession, it offered a greater margin of safety than orthodox methods then in use (for example, bloodletting, leeching, and purging). Despite the fact that the technology was crude (one had to make one's own battery and instruments for administering and measuring the currents), clinical reports by respected physicians indicated useful effects, particularly in growth control and psychiatric conditions.

However, by the late 1800s the field had become infested with quacks and charlatans. By 1900, it was almost totally dominated by this type of entrepreneur. A wide variety of fascinating devices was promoted, including electrical, electrostatic, and electromagnetic devices and simple permanent magnets. These had an alleged ability to cure everything from baldness to flat feet. None of the claims was in any way substantiated, and this circumstance reflected badly upon the few serious attempts to evaluate either the basic biological effects or the authentic clinical results.

During this time, vitalism was losing its scientific basis. Electricity was no longer viewed as the "life force," but simply as a form of energy that had useful clinical effects on its own. It was thought of as something akin to a drug, and if a little was good, more was considered to be better. Many respectable physicians devised therapeutic devices. They put forth elaborate theories, which were unsupported by fact, to explain these devices. The amounts of current/

voltage or field strength and frequency applied were based simply on the physician's intuition, and the placebo effect was ignored.

As scientific medicine became more firmly established, most such devices were shown to be of little value and totally without scientific foundation. By the 1930s some of them had disappeared entirely, although a surprising number persisted in use, with a steady trickle of patients being treated by the physician-inventors. At that time, the only scientifically valid use for electromagnetism in medicine was in the field of diagnosis—specifically, through the use of the electrocardiogram and the electroencephalogram, and of stimulator-type devices for the testing of nerve and muscle injuries.

Initially, the FDA felt no need to move against any diagnostic electrical device, and only the most overt therapeutic quacks were prosecuted under the agency's general rules of efficacy. By 1950, most of the fraudulent devices had been removed from the market. However, some of the devices promoted by legitimate physicians were left untouched on the premise that a physician may treat a patient in any way that he deems best. Until 1978, in fact, there were no specific rules governing medical devices, either therapeutic or diagnostic. Under the law enacted in that year, new devices of all types had to be approved by the FDA, but any devices that had been in clinical use before that time were "grandfathered"—that is, they were granted immunity from the regulation. These older devices could continue to be manufactured and used regardless of their efficacy or potential side effects. In actuality, the FDA appeared to adopt the position that electrotherapy was harmless nonsense and, if left alone, would soon disappear in favor of proven efficacious therapies. This simply did not happen. As the scientific revolution that began in the 1960s reintroduced electricity back into biology, the use of these grandfathered devices expanded, and new devices appeared.

THE RISE OF MODERN ELECTROTHERAPY

Before we discuss treatments involving the administration of electrical current to patients, some basic aspects of this process must be understood. For any current to flow, a complete electrical circuit has to be established. This requires that two electrodes be used: both may be on the body surface or inserted into the body, or one may be inserted and the other placed on the body surface. The current flows between the two electrodes, generally along the shortest, most direct route. Current or voltage density—that is, the amount passing through a defined unit area—is greatest near the electrodes. Along the flow path between the electrodes, the current will spread out, and the average current density will be less. In this flow of current, certain harmful physical events can occur.

First of all, electricity passing through any substance may produce heat. The amount of heat generated depends upon the resistance of the substance and the density of current flow. For human skin, the rule of thumb is that one milliampere (1 mA) per one square centimeter is just below the level at which cell damage due to heat is produced. Currents in excess of this may not produce perceptible heating, but they may still damage cells.

Second, if the electricity is administered via electrodes made of metal, the positive electrode will give off ions of the metal itself. In many cases, such as with stainless-steel electrodes, these ions will be quite toxic to cells. This toxic effect is not limited to the cells in direct contact with the electrode, because the positively charged ions will be electrically repelled by the positive field of the electrode and driven some distance deeper into the body.

Third, with any level of voltage, the water within the tissues will be subjected to electrolysis, a process in which the water molecules are broken apart. In biological tissues, this produces gases, such as hydrogen, that are extremely toxic to cells. The rate at which these gases are produced is proportional to the level of voltage: the higher the voltage, the greater the amount of gas. At low voltages, the circulation of blood and tissue fluids will carry the hydrogen gas away and neutralize it; however, as voltage is increased, a certain level will be reached at which gas production is greater than the ability of the circulation to carry it away. This higher voltage is called the electrolysis level, and it results in the rapid accumulation of hydrogen gas and the death of all cells at the site. In my laboratory we measured the onset of electrolysis in animal tissues using stainless-steel electrodes. We found it to occur at about 1.1 volts—a comparatively low level.

It should be apparent that electrical currents administered to the body may have purely physical side effects of an undesirable nature. In addition, as indicated earlier in this book, extremely small electrical currents have a variety of major biological effects. Most people who employ or promote electrical currents for treatments are either unaware of these effects or choose to ignore them.

THE "BLACK BOX" TREATMENT: TRANSCUTANEOUS ELECTRICAL NERVE STIMULATION (TENS)

Pain has been a puzzle to both clinicians and neurophysiologists for many years. While "tracts" of nerve fibers that transmit pain sensations had been identified in the spinal cord more than 100 years ago, the exact structures that initially detect the injury—the so-called pain receptors—remained a mystery. Clinically, pain is a subjective symptom that cannot be objectively measured. One had either to take the patient's word for it or prove the existence of some pathology that would justify the pain. Chronic pain, in particular, was a real problem. If the physician could not identify its cause and correct it, relief would require a long period of drug therapy, with all the attendant side effects.

In 1965, the gate theory was proposed as a technique for the relief of chronic pain. This attractive hypothesis proposed that pain signals from the periphery of the body to the brain could be blocked through the insertion of nonpainful electrical impulses into the dorsal columns (the spinal-cord tracts that carry the pain sensation). It was theorized that the gateway from these tracts into the higher centers of consciousness could carry only so much informational traffic, and that strong, nonpainful signals could block painful impulses.

Two years later, Dr. C. Norman Shealy, a neurosurgeon, reported that inserting stimulating electrodes into the dorsal columns produced relief of chronic pain in some patients. While this appeared to substantiate the earlier theory, Shealy's technique involved major neurological surgery, and it did not always work. In an attempt to determine in advance of surgery which patients would respond well to the electrode implantation and electrical stimulation, Shealy and his colleague, D. M. Long, applied electrical stimulation through skin electrodes over or near the area of pain. They theorized that stimulation of the sensory nerve fibers in this fashion would "block the gate" in the same way that electrodes implanted in the spinal cord did.

Initially, if patients reported control of pain with this technique, Shealy would proceed with the surgery and electrode implantation. Later, it became evident that the simple skin-electrode technique worked just as well as the surgically implanted electrodes, and the

surgical procedure was discarded as unnecessary. Several companies were formed to exploit this technology, and new, more efficient
devices were designed and marketed. This technique became known
as transcutaneous electrical nerve stimulation, or TENS. Because it
was in use prior to the 1978 law, it was a grandfathered therapy.

The actual technique involved the use of two electrodes, which
were attached to the skin over the painful area, and a "black box"
(a small, battery-powered pulse generator). The first generation of
devices, designed in accord with the gate theory, used high-frequency, low-current-density electrical pulses that neurophysiologists had determined were maximally stimulating to the sensory
nerves.

However, the gate theory was soon challenged by other researchers, and TENS units that used electrical pulses of low frequency and high currents and that did not stimulate the sensory
nerves were found to be equally effective in relieving chronic pain.
Another approach was to provide the patient with an adjustable
black box. In this way the patient could choose the pulse shape,
width, and frequency that seemed to provide the most pain relief.

It is now evident that a wide variety of waveforms, pulse
shapes, frequencies, and current densities will give considerable
pain relief; therefore, the effect cannot be due simply to the stimulation of the nonpainful sensory nerves. The gate theory has thus been
discarded. However, even though no mechanism of action can be
proved for their use, TENS units often provide considerable relief
of chronic, nonresponsive pain, and they have become an accepted
item in the modern physician's armamentarium.

I have studied a number of such devices and have found that
each one will eventually produce skin irritation at the electrode sites.
This side effect can be eliminated temporarily by switching to another type of unit, but eventually the same side effect will occur.
While I have found that these devices often produce considerable
pain relief, I do not believe that the potential side effects of high
levels of pulsating current have been fully evaluated. I have therefore stopped using the devices, and I do not recommend their use
unless all other possible techniques have been tried.

The TENS units have actually provided an umbrella covering
many other devices that use pulsing electrical currents and skin
electrodes. At this time, one can choose from a wide variety of
devices, all referred to as TENS units and all marketed for pain
relief. However, the actual clinical applications of these devices have

gone far beyond simple pain relief. Using a loophole in the FDA regulations that permits a registered physician to use an approved device in any fashion for any clinical condition on an individual patient, manufacturers and physicians have employed TENS units for a multitude of nonrecommended applications. They have, for example, been used to treat migraine, with the electrodes applied directly to the head; a similar application has been made to treat drug addiction and memory loss. They are also being used to increase the rate of healing of wounds in both soft tissue and bone and to treat bladder dysfunction, as well as to serve as electrical stimulators in acupuncture treatment.

Major use has been made of TENS units in sports medicine, where it is claimed that they reduce edema, dissolve hematomas, and restore full function to injured parts. Military jet-fighter pilots apply the electrodes in advance of getting into cramped cockpits for extended periods of time, claiming that doing so prevents backache. It is obvious that most of the above applications have little to do with chronic pain, and I believe that most have no present basis in either scientific theory or research.

We have come full circle, from widespread unregulated use of electrical-stimulation devices, to the total discreditation of all electrical effects, and now back to a widespread acceptance of such devices for the treatment of a multitude of conditions. There are several unfortunate aspects to this development. First, the original theoretical foundation for their initial use—the gate theory—is no longer accepted, so the mechanism by which these devices reduce pain is unknown. Second, at present there is no scientific basis for their use in non–pain-relieving applications. Third, there has been no recognition of, or search for, any deleterious side effects. This latter aspect is particularly important: each TENS unit exposes large volumes of a patient's tissues to substantial electrical currents that are pulsing at frequencies not normally present in the Earth's electromagnetic field. As we shall see, such electrical forces do have important, and often very undesirable, side effects.

There are so many devices currently on the market masquerading as TENS units that it is impossible to review the characteristics, theoretical foundations, and proposed applications of each one. Suffice it to say that at this time, my recommendation would be that the use of these devices be limited to chronic painful conditions that are unrelieved by standard treatments, and their application limited to areas of the body other than the head, neck, and spinal cord. Each patient must be thoroughly evaluated in an attempt to determine the cause of the pain so that other treatment methods can be tried first.

ELECTROTHERAPY FOR DRUG ADDICTION

The idea that electrical stimulation to the head could be beneficial in the treatment of drug addiction was derived primarily from the work of Dr. Margaret Patterson, a British surgeon. About fifteen years ago she came to my laboratory with an interesting story. While she was completing her surgical training at a British hospital in Hong Kong, a Chinese surgeon had shown her that auricular acupuncture (a technique that uses points on the patient's ear) could be used in place of the usual large doses of narcotics to prevent withdrawal syndrome in postoperative drug addicts. This technique so impressed Dr. Patterson that she began to use it on her patients there. She then returned to England, where she built up a busy surgical practice. Later, with the increase of drug addiction in England, Dr. Patterson began using postoperative acupuncture on those cases. Her fame spread, and patients began to consult her primarily for acupuncture treatment of drug addiction.

When she learned of the Chinese use of electrical stimulation to enhance the effect of acupuncture, Dr. Patterson adopted that technique, using pulsed DC. However, she encountered problems. While the clinical results were better, there was considerable local irritation from the current, and it was difficult to keep the needles in place in the ear. Nevertheless, she believed that this was a useful treatment for drug addiction and that it was better than any other, and she began seeking a method that would obviate these difficulties.

When she visited my laboratory, Dr. Patterson told me of these problems and asked if I could suggest a way to improve the technique. My sole contribution was to recommend that she stop using needle electrodes and instead use a flat, surface electrode of at least one square centimeter in size, applied to the skin just behind the ear. I told her I thought that only about 1 percent of the total current she was applying to the needle electrodes actually traversed the brain, but that this appeared to be enough to do the job. It seemed to me that the effects of her electrical acupuncture treatment were probably not mediated by the acupuncture system at all but were somehow due to the electrical current's acting directly on the brain. I suggested that she explore various frequencies to determine whether any were more effective than others, and also whether any could be related to addiction to a specific drug.

She followed this advice, and since then we have kept in close contact. A few years ago I visited England and examined a number of Dr. Patterson's patients. What I found was amazing. Even severely addicted patients could completely stop all drugs as soon as the electrodes were applied, and they showed no sign of withdrawal symptoms. However, I was most impressed by the fact that every patient reported having experienced a major change in personality during the six-week treatment: all of them believed that during the treatment time they had gone from being addictive to nonaddictive personality types.

In animal experiments, Dr. Patterson had found that during electrical treatment the amount of the brain's endorphins (a natural, morphinelike substance produced by the brain) increases measurably. While this may explain some of the immediate results, it cannot be the only answer, because the majority of her patients require only one six-week treatment and remain drug-free thereafter. It seems likely that there is some long-term effect that persists after treatment is stopped.

There are several interesting aspects to this technique. First, it seems that very low levels of pulsed electrical current have major effects upon the highest functions of the brain. The personality alteration that follows the treatment is a most significant observation that urgently requires further study. Second, from all accounts this treatment is so superior to others for drug addiction that it certainly deserves a large-scale scientific study to determine its actual clinical utility.

From 1944 to 1950, when I worked at Bellevue Hospital in New York City, one of the largest and busiest hospitals in the United States, I saw very few drug addicts, despite the fact that this was a time of great social and economic stress. I need not indicate the seriousness of the drug problem today. Yet, to the best of my knowledge, not one study has been done to address the question of the cause of the increased demand for addictive drugs. Instead, we have viewed the problem as being produced by an increased supply, and we have therefore based our countermeasures on interdicting the supply of illicit drugs. Unlike other treatment methods, Dr. Patterson's seems capable of actually decreasing the *demand* for the drugs. It has been used with great success in the treatment of prominent British rock stars, but it has not yet been approved for the treatment of the average citizen. Dr. Patterson's discussions with the NIH and the FDA in this country have, so far, been unfruitful.

While I generally take the position that all electromagnetic therapy has a dark side and should be used with great caution, I also believe in risk/benefit analysis. If the condition for which the treatment is to be employed is life-threatening, if no other therapies have the same level of effectiveness, and if the risks of producing undesirable side effects are less than those of no treatment at all, then the use of electromagnetic therapy is justified. The use of Dr. Patterson's treatment for severe drug addiction would seem to be justified by this risk/benefit ratio concept. At the least, a large-scale clinical study should be undertaken to either prove or disprove her observations.

A large number of addicts are currently being treated, at a high cost, by a variety of people who are using different types of TENS units (or other devices, including some that produce pulsing magnetic fields) applied to the head. Not only is this unscientific, but it is also dangerous.

There is an even darker side to this situation. If we can treat drug addiction by passing an electrical current through the head or exposing the head to a strong, pulsing magnetic field, then we should be able to mimic the effects of certain illicit drugs, such as psychedelics or opiates, using the same technique. This idea has in fact become common, and it has been used in a technique called, in the drug culture, "wire-heading." The plans for these kinds of devices circulate freely, and any electronics store can provide the materials to construct a simple, but powerful, pulse generator for this purpose. The practice is not strictly illegal, and I doubt that it can be controlled in any legal fashion. It is, however, decidedly dangerous.

A year or so ago, I received a call from a gentleman who works in the music industry, and who had worked with a young man who had formerly been a serious drug addict. The two had discussed the possibility of using electronic devices to mimic drug effects. The former addict believed that the use of such devices would be safe, because they would not produce a dependency response. Music studios use multitrack magnetic tape for recordings. In order for the tape to be reused, it must be degaussed—that is, placed in the field of a powerful electromagnet oscillating at 60 Hz—to erase the prior recording.

The young man decided to test his theory, and he applied degaussing coils to each side of his head. Over the next hour he became increasingly psychologically disturbed and had to be hospitalized that evening. He was not able to return to work for several months, and by that time he had lost all memory of the incident. I do not know if a similar reaction would occur in a subject who had not previously been seriously addicted, but I do *not* recommend anyone's doing this, or any related, experiment.

ELECTRICITY AND THE HEALING OF BONE FRACTURES

The FDA's approval in 1978 of several types of electrical and electromagnetic devices for stimulating the healing of human bone fractures that have failed to heal has played a major role in gaining acceptance for electrotherapeutics in general. More than 100,000 patients have been treated in this country and abroad with these devices. The industry has continued to produce new types of devices, and it has been aggressively searching for an expanded market in treatments for other conditions. The way in which this all began is an interesting and instructive story.

In 1960, I presented my data on the electrical factors related to salamander-limb regeneration at the annual meeting of the Orthopedic Research Society. After my presentation, Dr. Andrew Bassett, a young orthopedic surgeon from the Columbia University College of Physicians and Surgeons, introduced himself to me and suggested that we work together on possible electrical factors in bones and in bone healing. We began a joint study within a few months, with our first project being a search for a piezoelectric effect in bone.

The piezoelectric effect is the generation of an electrical charge in some crystalline materials, produced by bending or pressing on these materials. The charge is produced instantaneously and rapidly diminishes to zero. When the bending or pressure is released, the same amount of instantaneous charge is produced in the opposite direction. Piezoelectric crystals are frequently used to produce sparks to light propane stoves and other similar devices.

Bone has three types of growth: the increase in length and thickness during childhood, the healing of fractures, and the response to mechanical stress. This third type indicates that bone is somehow able to "sense" mechanical stress and grow, so as to produce an anatomical structure best able to resist the stress. Because bone is composed of two different materials—a protein fiber called collagen and a mineral crystal called apatite—we theorized that the mineral crystals might be piezoelectric, thereby providing the ability to sense mechanical stress.

After some investigating we were able to show that bone is capable of producing the piezoelectric effect. However, the two charges (from bending and releasing) were not equal, suggesting that another factor was operating to limit, or rectify, the current flow in one direction.

Electronic devices called "rectifiers" or "diodes," which are based on single crystals of various types, pass electrical current in one direction only. If an alternating current or a series of alternating pulses (as from a piezoelectric crystal) is applied to the input terminal of a diode, only those portions of the current going in one direction will come out of the output terminal. Not all diodes are equally efficient; some pass a part of the current going in the reverse direction. This was what seemed to be happening in bone that was being repeatedly stressed. However, one might not expect a biological

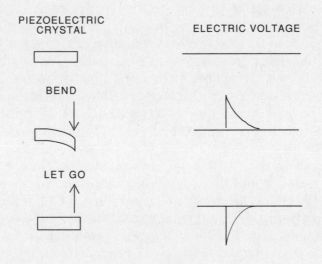

FIGURE 6-1. *Bending a piezoelectric crystal produces a very short surge in voltage; releasing it produces exactly the same surge in voltage, but in the opposite direction or polarity.*

system composed of millions of tiny crystals to work quite as well as a large, single crystal.

By 1964 we were able to describe the control system regulating bone growth that occurs in response to mechanical stress and that operates on the basis of the piezoelectric effect. It is important that the piezoelectric effect in bone be partially rectified and produce a signal that has more current going in one direction. This imbalance in the electrical output provides the direction in which new bone must be built. If the effect were the same as that from a single large crystal going equally in both directions, there would be no directionality, and it would be useless as a control-system signal.

Stress-related bone growth is actually remodeling of the shape of a bone so that it can better resist mechanical stress. The piezoelectric effect "tells" the bone how much stress is being applied and in what direction. The mechanisms associated with this type of bone growth and the cellular events that bring it about are totally different from those involved in fracture healing. I make a point of this because many people (a majority of orthopedic surgeons included) have confused the two.

Turning On Bone Growth Electrically

In the same year, we began experiments on stimulating bone growth using direct electrical current, which is very different from the short, transient pulses associated with the piezoelectric effect. The idea of using DC rather than piezoelectric-type pulses was based on my original observation that negative DC current was required for regeneration in the salamander. By then Bassett had acquired a skilled electrical technician, Robert Pawluk, who made the devices we used. These were small (about half an inch long), battery-powered units with two platinum electrodes each. The units were implanted so that both electrodes entered the marrow cavity of intact, nonfractured thighbones in adult dogs.

When Pawluk asked me how much current the devices should produce, I had to say that I really did not know, so we had to make some educated guesses. We settled on three different current levels: 1, 10, and 100 microamperes. In this way we hoped to cover the possible range of effective values. Pawluk measured the current from each unit for about thirty minutes following its implantation in the dog's leg. Over that short period of time, he found that the

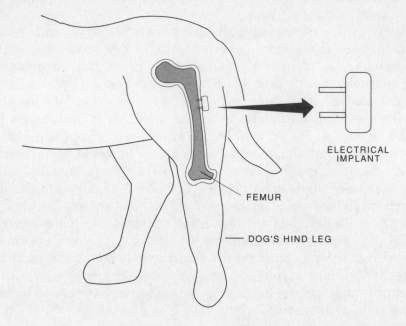

ELECTRICAL
IMPLANT

FEMUR

DOG'S HIND LEG

FIGURE 6-2. *Implantation of electrical stimulator in dog's femur. The two electrodes, positive and negative, project into the marrow cavity of the bone.*

current from each device decreased substantially, so that in reality we were using about ¼, 1, and 3 microamperes maximum from the three types of units.

The experiment ran for twenty-one days. At the end we found that a large mass of bone had grown in the marrow cavity around the negative electrode at 3 microamps and a slightly smaller amount had grown around the 1-microamp negative electrode; no difference was found between the ¼-microamp negative electrode and simple control electrodes with no current. There was some apparent loss of bone in the vicinity of the positive electrodes, but this was difficult to quantify.

As we neared the end of both the piezoelectric and DC bone-growth projects, we discovered that both experiments had been done before (although not in exactly the same manner) by several Japanese scientists. Our ideas were, therefore, not new, and our data had been anticipated by the work of other scientists. Nevertheless, our two reports stimulated great interest in the orthopedic community.

Unfortunately, however, our work was widely misinterpreted. Little attention was paid to its most important aspect, which was that for the first time in the history of biology or medicine, a normal growth process had been deliberately "turned on." Until then, nature alone had held the key. Most people also confused the production of bone growth by negative DC with the pulsed piezoelectric effect, and they thus concluded that bone could be electrically stimulated simply because it was piezoelectric. This was interpreted to mean that the application of these electrical currents in the body would have an effect *only* on bone, not on any other tissue. The fact that we had not stimulated bone growth itself but had actually produced an activation of bone-marrow cells and their subsequent transformation into bone was ignored. In retrospect, I see that we were in part responsible for this, because we did not emphasize these points and indicate their significance.

The result of the simplistic interpretation of our work was the popular idea that bone was unique in its response to electrical factors, that negative electrodes stimulated bone growth in some "magical" way, and that the effective level of current was about 10 microamperes. Some orthopedic surgeons seemed to think that these data were sufficient for them to begin clinical usage of the technique.

I did not agree with this view. In my laboratory we concentrated on the basic aspects of the relationship of electrical forces with *all* types of growth, as well as with other physiological systems (particularly the nervous system). In 1970, we were able to report the details of the complete DC electrical control system that governed fracture healing in frogs.

The scientific literature in 1970 indicated that frogs healed their fractures in the same fashion as humans: by cell division, which produced new bone-forming cells. However, we found that this was not quite correct for either frogs or humans. We were able to show that fracture healing is basically a regenerative process that involves the electrical dedifferentiation of certain cells.

In frogs, the red blood cells in the blood clot at the fracture site are dedifferentiated by the DC electrical current and subsequently redifferentiated into bone-forming cells. (The red blood cells of the frog still contain their nucleus, so they still have all their genetic information and can dedifferentiate. Our red cells do not have a nucleus, hence they cannot dedifferentiate.) In the human being, the cells of the bone marrow are also electrically dedifferentiated and subsequently redifferentiated as bone cells. Actually, this was the process that Bassett and I had experimentally produced in our first experiment with dogs. We were able to describe the details of this cellular process and show that it is produced *only* by extremely small levels of voltage and current. In this case, more is definitely not better.

We also were able to show that a frog's red blood cells can be electrically dedifferentiated in culture dishes, and that they have the biochemical markers for dedifferentiated "turned-on" cells. Most importantly, we were able to show that the electrical control signal responsible for the bone-healing process begins at the time of the fracture, runs for a fairly definite period of time, and then declines to zero. The time sequence of this negative DC current is similar to that in the regenerating salamander limb.

This is an important observation because it explains why fractures sometimes fail to heal, resulting in what orthopedic surgeons call a nonunion. The DC current produced at the fracture stimulates the dedifferentiation in cells of human bone marrow just as it does in the cells in the salamander's limb or in the red blood cells in the frog's fracture hematoma. When this DC current returns to zero, cellular stimulation ceases, and healing stops.

If the two ends of a fractured bone in a human are separated or are permitted to move, the cellular healing process—which still

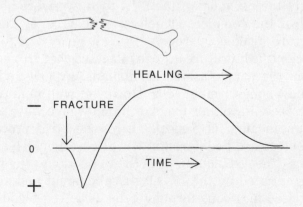

FIGURE 6-3. *The electrical-voltage changes immediately following the fracture of a frog's bone (tibia) and during healing.*

proceeds normally at each end of the broken bone—will not be able to make a connection across the fracture gap. If the bone has not connected across the fracture gap when the stimulating DC electrical current has run its course and dropped to zero, the result is a nonunion. Surgery and bone-grafting procedures actually restart this electrical control system, because they injure the bone and re-stimulate the DC electrical control system. Therefore, to stimulate the natural DC fracture-healing system with externally generated electrical forces, one has to apply the appropriate amount of negative DC electrical current to the cells of the bone marrow at the fracture site.

Our colleagues in the orthopedic community were less inter-ested in this type of basic work than in bringing this technology to clinical fruition. In 1971, Dr. Zachary Friedenberg of the University of Pennsylvania reported the first successful electrical treatment of a nonunion of a fracture in a human. He used 10 microamps of DC current, with a stainless-steel negative electrode imbedded directly in the bone marrow at the fracture site (see Figure 2-5, chapter 2). This promising beginning was rapidly expanded by Dr. Carl Brigh-ton, Dr. Friedenberg's colleague, through the establishment of a clinical-research program based upon the same technique. Shortly thereafter, Bassett and his colleagues proposed an alternate tech-nique that did not require surgery and that was based on the induc-tion of electrical currents in the bone by means of an external pulsing magnetic field.

I was certain that a pulsed electromagnetic field (PEMF) would simply produce a pulsed, alternating electrical current in the bone, and that this could not stimulate the bone-marrow cells to dedif-ferentiate. Furthermore, the pulsing magnetic field itself would per-meate the tissues around the fracture site, and its own biological effect might be considerable. Nevertheless, Bassett and his cowork-ers thought that the use of an unbalanced magnetic pulse would produce the equivalent of a pulsing DC field in which the bone was primarily exposed to a negatively charged electrical environment. The possibility that the magnetic field had any biological action was simply discarded. Their animal experiments seemed to validate this idea; fresh fractures exposed to the pulsed magnetic field healed significantly faster than normal.

The healing time was determined by exposing the bones to mechanical stress at various times. If the fracture was healed, the amount of mechanical stress required to break it was within the

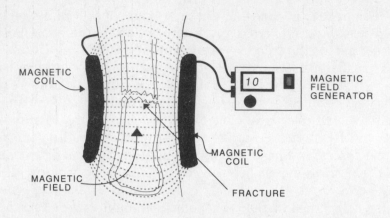

FIGURE 6-4. *The pulsed-electromagnetic-field treatment for bone fractures that have failed to heal. Magnetic coils are used on each side of the leg, generating a magnetic field that penetrates all of the tissues, including the fracture. As this field is pulsed, it generates a pulsing electrical current.*

range of that for nonfractured bones. While this is a very accurate method for determining the time of *mechanical* healing, it tells us nothing about the *cellular* mechanisms involved.

Following FDA approval, these techniques were used clinically. Despite obvious differences in technique, they were found to produce identical success rates of about 80 percent in inducing bone growth in human patients.

I was of the opinion that the available scientific data were insufficient to warrant FDA approval of these devices. None of the proponents of any device had described the cellular responses produced by the particular technique involved. The 10-microampere level of current in the DC devices had been assumed to be optimal. As a result, no other current levels had been investigated, no one had determined whether electrolysis was being produced, and the potential biological effects of the pulsed electromagnetic field had not been assessed in the PEMF device.

I decided to try to accurately measure the level of voltage at which overt electrolysis occurred in mammalian tissues. We found that this varied with the type of metal used for the electrode; however, the average electrolysis voltage was disturbingly low. For example, the stainless-steel electrodes used clinically started producing gas at 1.1 volts of DC current. The current level that was used clinically, 10 microamperes, required voltages well above this level. This meant that the devices in clinical use were producing damage and even death in cells at the treatment site. I postulated that the mechanism of action in these devices was the production of a local injury, with reactive bone growth then occurring in response to the injury. This certainly was not a biologically desirable action, and it was unrelated to restarting the normal healing process when this had failed.

Our measurements of healing fractures in animals indicated that much smaller voltages and currents were used by the body's growth-control system. I believed that we should be able to use these same electrical values and produce bone healing without damage. In a series of clinical cases, we were able to show that the same rate of success could be obtained using silver electrodes with currents a hundred times lower than those in use, and with voltages below electrolysis levels. Instead of 10 microamperes, we used 0.1 microampere, and we proved that this minute current/voltage level would still cause bone growth and healing of the nonunion. The silver electrodes were important, because their low resistance permitted the use of low voltages for passing the desired current level. I believe that with this technique, we were actually turning on the growth-control system, rather than overloading the system with far too much energy and thus producing damage.

LINKING ELECTRICAL CURRENTS AND CANCER GROWTH

It seemed to me that any electrical technique capable of stimulating bone growth might also stimulate other types of growth. In 1981, I experimented with human cancer cells in culture, exposing them to levels of DC current equivalent to those present in the area of soft tissue outside of a bone nonunion treated with the 10-microampere technique. The current flow from the implanted electrode to the skin electrode exposes large volumes of tissue to currents and voltages that are below those producing electrolysis, but that are still capable of causing the turning on of a growth process. *The cancer cells exposed to these electrical factors grew at least 300 percent faster than the controls.* I was surprised to find that this significant increase in growth occurred at both the negative and the positive electrodes. All of the preceding experiments had shown that growth was associated with negative polarity and the cessation of growth with positive polarity.

Until I did this experiment, I had believed that cancer cells might increase their growth rate with negative electricity, but that positive electricity might slow or even stop their growth. This experiment showed that cancer cells were really very different from normal cells. Further, it showed that the DC current devices, approved by the FDA, were promoters of cancer growth. I reported my results in one of the orthopedic journals, pointing out that the approved DC current technique might stimulate the growth of an unsuspected cancer anywhere within the current pathway.

In 1981, a team of Japanese investigators reported that mouse bone-cancer cells treated with the same DC technique increased their DNA synthesis (a measure of cell multiplication) by about 200 percent, substantiating my results.

How Pulsed Electromagnetic Fields (PEMFs) Work

In 1983, Dr. Abraham Liboff, a professor of physics at Oakland University, reported that the effect of pulsed magnetic fields on normal cells in mitosis was an increase in the synthesis of DNA. Furthermore, he said that this effect was due to the magnetic field and not to any induced electrical currents. Experiments in other laboratories quickly confirmed Liboff's observations. The effect was clearly due to the magnetic field, and it occurred only in cells in active growth. This explained why the clinical reports of the efficacy of the PEMF technique varied so greatly: some clinicians reported the usual 80 percent success rate, while others reported almost a zero rate.

The success rate depended upon when the clinician declared a fracture to be a nonunion. If this decision was reached early in the clinical course (at three months after the fracture, for example), active cell growth was still occurring that could be stimulated by the pulsed magnetic field. If the clinician was conservative and waited until any chance of normal healing had ceased, there was no active cellular multiplication at the fracture site, and the pulsed field would have no effect. The mechanism of growth stimulation by pulsed magnetic fields was a *magnetic effect on cells in active mitosis*, not a turning on of the normal DC growth-control system.

Also in 1983, a Japanese investigator using an actual therapeutic PEMF device reported a major increase in the growth of cancer cells in culture, as well as an increase in the malignant characteristics of these cells. Cancer cells in mitosis were more sensitive to magnetic fields than normal cells.

In response to these reports, I wrote to the FDA and suggested that the agency reappraise its present approval. The FDA representatives who replied indicated that they were aware of the reports; however, because the four-year follow-up studies on patients treated with these devices had not been completed, they considered the data base to be incomplete, and they took no action.

Since then, I have consulted with several medical-device manufacturers who were preparing other electrical bone-growth–stimulating devices for submission to FDA. The data each manufacturer was required to submit included a report of the effect of the device on cancer cells in culture. All of these manufacturers reported a significant increase in cancer-cell growth. Nevertheless, FDA granted approval for clinical testing to prove efficacy only.

Some years ago, I expressed the opinion that overenthusiastic application of the electrical and electromagnetic bone-growth stimulation techniques could lead to disaster if it resulted in harmful side effects. Such an event would prevent acceptance of any electromagnetic therapy in the future. One can only hope that the present circumstance will not prove me correct. The lesson still to be learned is that we must work with nature rather than trying to improve on it.

THE HISTORY OF ELECTROTHERAPY AND CANCER

Exactly when electricity was first used as a possible treatment for cancer is unknown. The quotation from Dr. A. Allison at the beginning of this chapter is taken from his letter to the *Lancet* of January 10, 1880, in which he described a case he had treated. His patient was an English farmer who had developed a cancer of the lower lip and chin. The patient had agreed to have surgery performed on the lesion, but before it could be performed he was struck by lightning while working in the fields. Dr. Allison was called and found the patient in "a state of great prostration." When the farmer regained his senses, the doctor "bled him from the arm." In his letter, Dr. Allison went on to state,

What seems to be the most astonishing feature in the case is the healing process which was set up in the lip and chin soon after the accident. The cancer gradually lessened, and in a few weeks every trace of the diseased structure gradually disappeared, and for ten years he enjoyed complete freedom from his former suffering and signs of the disease.

In reading this account more than a hundred years later, I believe that the most significant aspect of the case was the *gradual* disappearance of the cancer. One must assume that Dr. Allison was a careful observer and, therefore, that the cancer was not killed by a direct action of the electrical discharge, which would have turned it black and necrotic. The gradual disappearance would seem to indicate the operation of another factor, one that was set in motion by the severe electrical injury. To my mind, Dr. Allison's plea for "testing to the fullest extent the curative effects of frictional electricity" remains well founded.

To the best of my knowledge, electricity was first used for this purpose in a scientific manner later in the 1880s by Professor Apostoli, a French surgeon. Even though anesthesia had been discovered in 1848, surgical techniques had lagged behind. And in the 1880s, nonsurgical techniques were still preferred for treating cancers of the cervix and uterus. Apostoli treated these types of tumors with DC electricity, inserting the positive electrode into the tumor and passing between 100 and 250 milliamperes (mA) of current through the tumor to a large negative electrode on the abdomen (clearly, he was producing electrolysis within the tumor). He reported prompt relief of pain and bleeding, and shrinkage of the tumors, but he reported no long-term results.

Sometime around 1960 I was given an old book, the autobiography of Dr. Franklin H. Martin. Martin had been a young surgeon in the 1880s and had gone on to considerable prominence in American surgery. He had become interested in Apostoli's method of treatment, and in the book he described his experiences in using this technique on his patients. Martin was able to confirm Apostoli's results. He published several papers on the method, acquiring a certain notoriety in the process.

A few years later, according to Martin, "There [was] intimation that the use of electricity in the region of the pelvis would destroy the fertile ovum, or the embryo in the process of development . . . [and] that careless use . . . would terminate pregnancy." Martin characteristically decided to look into this. Using fertile chicken eggs, he found that a direct current of only 20 mA destroyed the chick embryos. He therefore immediately discarded this method of treatment.

Martin became caught up in the subsequent remarkable development of surgery, and by 1900 he was involved in devising new and radical surgical techniques for cancer and in establishing the American College of Surgeons, as well as some major medical journals. The only further references to work on electrotherapeutics in his book are remarks decrying the entry of quackery into this field, which he still considered promising.

MODERN ELECTROTHERAPY FOR CANCER

As knowledge of the physics of electricity increased and better methods for generating and measuring it were developed, all rapidly growing tissues were found to be negative in polarity compared with the rest of the body. The highest negativity was found in malignant tumors. Based on these observations, doctors Carroll Humphrey and E. H. Seal of the Johns Hopkins Applied Physics Laboratory postulated that cancer growth could be speeded up by negative currents and slowed down by positive currents. Their report, published in 1959, seemed to support this idea.

Their study involved mice that had had a rapidly growing cancer implanted just beneath the skin of the back. The DC electrical application was begun twenty-four hours after implantation (the cancer cells were just beginning to grow at this time). Humphrey and Seal used surface electrodes of either copper or zinc, applied directly over the cancer-implant site, with currents far lower than those that had been used by Martin. They found that they could not make the cancers grow any faster than usual with negative electrodes, but they did report a significant growth retardation when positive electrodes were applied over the tumor.

While the surface electrodes did not produce electrolysis, ions of copper or zinc were released from each positive electrode and were driven into the tumor cells by the voltage. Copper and zinc are both known to be toxic to cells, and this is most likely the mechanism involved in the results obtained by Humphrey and Seal. The experiment would have been clearer if the tumor cells had been allowed to grow for several more days, producing a noticeable mass. Humphrey and Seal suggested that continuation of the studies with other types of tumors would be desirable, but apparently they never did this follow-up themselves.

In 1977, doctors Muriel Schaubel and Mutaz Habal of the State University of New York Upstate Medical Center reported on basically the same type of study, using rats with implanted tumors. In this study, however, the treatment electrodes were stainless-steel needles inserted directly into the tumors. Schaubel and Habal used

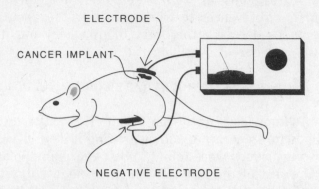

FIGURE 6-5. *Experimental electrical control of cancer. The cancer cells are implanted just beneath the skin of the back. Twenty-four hours later, the positive electrode is placed over the cancer implant, and the negative electrode is placed on the skin of the abdomen. The cancer cells are then exposed to the positive electrical factors from the electrode.*

three levels of current: 3 milliamperes, ½ milliampere (500 microamps), and a third, much smaller, level of 960 millimicroamperes. They also studied the effects of both positive and negative currents on the tumors at each current level. With the 3-mA current there was significant destruction of the tumor at both positive and negative electrodes, with about twice as much at the positive as at the negative. At the ½-mA level there was also destruction of the tumor, but only with the positive electrode. At the lowest level of current there was a reduction in the weight of the tumors with both positive and negative electrodes, but there was no tumor destruction.

The number and size of tumors in other parts of the body, which had spread from the main tumor via the bloodstream, were reduced in the animals treated with 3 milliamperes of positive current but not in those treated with the negative current. However, a similar finding was noted with both the positive and negative treatments using the smallest level of current. Consistency would require that some similar effect would have been noted at the current levels between these highest and lowest currents; however, this was not the case.

At this time, the best conclusion that can be drawn from this experiment is that the tumor destruction was the result of local electrolysis at the needle electrode. The observations on the metastases are interesting but difficult to evaluate. The metastases were not directly exposed to the electrical currents, and we must therefore assume that effects on them were due to effects on the primary tumor. Tumor-necrosis factor was not known at the time, but it appears conceivable that it might have been produced by the necrosis of the primary tumor.

Nothing more was heard of this type of study until recently, when Professor Björn Nordenström, a radiologist at the Karolinska Institute in Sweden, published his book postulating "biologically closed electric circuits." Nordenström became interested in bioelectricity when he observed the structure of lung tumors on X-rays. The radiating lines extending out from such tumors reminded him of the electrical corona, and he began an investigation of many years' duration into the movements of electrical currents in the body. It is not possible to review here the entirety of Nordenström's work, which covers a large area of "bioelectricity." In essence, his basic concept of closed electrical circuits is complex but appears to have little support biologically or in the scientific literature.

Nordenström's cancer-treatment method differs little from those of Apostoli, Martin, or Schaubel and Habal. Nordenström inserts stainless-steel needle electrodes directly into lung tumors under X-ray visualization and applies 10 volts of positive electricity, with a negative electrode applied to the skin of the chest. Many of his published X-rays show a gas bubble, indicating electrolysis and general tissue destruction.

The reports of all of these researchers share the same defects due to lack of attention to basic principles of physics and electrochemistry. They are derived from the information that rapidly growing tissues are electronegative and from the simple concept that positive electricity should, in some magical way, cause a cessation of the tumor growth. In Nordenström's case, the theoretical construct was different, but the final approach was the same. However, we now know that the idea that positive electrical current inhibits cancer-cell growth is incorrect.

Cancer cells respond to both positive and negative electricity by growing rapidly. The degeneration of the tumors observed by Humphrey and Seal can be attributed to the fact that the current in the vicinity of a positive metallic electrode is carried by ions given off from the electrode itself. Ions of copper, zinc, and stainless steel (which is actually an alloy of many different metals, including cobalt) are toxic to all cells, and cancer cells are no exception.

In the higher current levels used by Apostoli, Martin, Schaubel and Habal, and Nordenström, electrolysis and gas formation occurs within the tumor. The result is a massive shift in the local acid-base balance, with the production of a highly acidic area within which cells are destroyed. In neither case is the destruction of the tumor the result of electrical factors acting alone or via some influence from an electrical-control system.

Despite these "theoretical" objections, the tumors *are* destroyed. If this is what we really want, why shouldn't these techniques be used? The reason is that apart from local toxicity, there is a very real hazard of stimulating other cancer growth with this use of electricity. As the current flows, it spreads out and flows along a return path to the negative electrode. In so doing, its strength (or concentration per unit volume) drops below the level of electrolysis. In this way, a large volume of tissue is exposed to electrical currents that stimulate rather than retard tumor growth.

In practicality, treatment of a single large tumor nodule will result in destruction of the center of the cancerous mass by electrolysis. But as the current spreads through the rest of the tumor, it will be below electrolysis levels and will cause growth stimulation of the outer portions of the cancer nodule. Also, if there are several cancer nodules present in the area, treatment of one may result in its necrosis, but—as occurred in several of Nordenström's cases—the spread of the current on its return path will stimulate the growth of adjacent tumors. Nordenström was unable to explain this effect.

Although modern studies on the relationship between simple electrical currents and cancer give us little hope for any effective therapy, Dr. Allison's early observation remains unexplained. It seems that we are missing something, and the answer appears to lie in a more sophisticated approach.

MAGNETIC-FIELD THERAPY FOR CANCER

The use of magnetic fields to treat cancer has a long and shady history, involving outright fraud along with the few serious attempts to evaluate the effect of such forces upon malignant growths. In the early years of this century, when science excluded the possible effect of electromagnetism on living things, this area of scientific research was apparently finally discredited. Now, with the new data on the biological nature of cancer and the relationship between magnetic fields and living organisms, we can no longer be so sure.

In the early 1960s, as my own work was progressing, I was

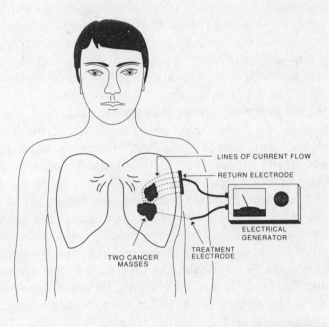

FIGURE 6-6. *The use of a positive electrical current to treat a cancer mass in the lung, with an adjacent cancer mass. The electrical current flows back to the return electrode on the skin. In so doing, it passes through the adjacent cancer mass, which is not being treated. Because the current/voltage in the second cancer is below the electrolysis level, the cancer cells in this tumor are stimulated to grow. While the first cancer is destroyed by electrolysis, the second cancer grows larger.*

contacted by Dr. Kenneth McLean, an M.D. practicing obstetrics and gynecology in New York City. Some years before, McLean had visited a mine in northern Canada, where he had been told that there had been no cancer among the miners for several generations. The miners attributed this phenomenon to the very strong steady-state, or DC, magnetic field of the mine. This gave McLean an idea, and he convinced some of his friends in industry to loan him some very powerful industrial electromagnets. After conducting animal experiments using these electromagnets, he claimed that rats inoculated with cancer cells survived if they were treated with extremely high-strength DC magnetic fields. In the late 1950s, he began using the technique on human patients.

Dr. McLean was much more of a clinician than a researcher. Being more interested in his clinical results, he often failed to set up well-controlled experiments or even to collect the appropriate data. In some cases, he obtained results that were truly extraordinary, yet the placebo effect could not be excluded. Unfortunately, the past history of "magnetotherapy" and the present dogmatic attitude have, as yet, prevented any scientific evaluation of McLean's ideas. We simply do not know whether he was correct.

In the past few years, pulsed-magnetic-field treatment for bone nonunions also has been reported to slow the growth of animal tumors, with a significant prolongation of survival time. This would appear to contradict those reports that pulsed magnetic fields produce marked enhancement of tumor-cell growth in culture. Was there, perhaps, some difference in response when the tumor cells were *in vivo* (that is, within the living organism)? The answer is no; the difference is in the way the experiment is done. In the culture studies, only the cancer cells were exposed to the pulsed magnetic field. In the animal experiments, the total animal was exposed to the pulsed magnetic field. Thus, there would be effects not only on the cancer cells in the animal but also on the animal's organs and systems that responded to the magnetic field.

As we shall see in the next chapter, time-varying, or pulsed, magnetic fields have a major effect on the stress-response system. Exposure of the whole animal for a short time causes a rapid stress response, with a marked increase in the activity of the immune system. Therefore, total body exposure, while stimulating the implanted cancer cells to grow faster, also enhances the action of the immune system. For a time, the immune system has the upper hand, and tumor growth is inhibited. However, continuing the exposure beyond the short term results in a decline of the stress response and the immune systems to below-normal levels. Tumor-cell growth is then enhanced by both the drop in immune-system efficiency and the direct action of the pulsed magnetic field on the cancer cells themselves.

Application of the PEMF technique to cancer treatment in humans would not (for technical reasons) entail total-body exposure. The pulsed field would be directed against the tumor itself, in which case a growth-stimulating effect on the cancer cells would occur.

None of the above should be taken as an indication that electrical or electromagnetic therapy for cancer is impossible. What it does indicate is that our current simplistic approaches are doomed to failure. As more is learned about the basic mechanisms responsible for the bioeffects of such fields or currents, it is quite probable that more sophisticated approaches will be developed. Some of the possible approaches will be discussed in chapter 10.

ELECTROCHEMICAL GROWTH STIMULATION

In the early 1970s, I began a study of the biological effects of different kinds of metallic electrodes. I have already mentioned that the current in the vicinity of a positive metal electrode is carried by means of ions of the metal itself. These ions are themselves "positive" because they lack one or two electrons. The positive voltage on the electrode repels the positive ions, and they are pushed, or driven, into the tissues. The result is a unique situation, with a voltage field containing large numbers of positive metal ions that can chemically react with membranes of living cells.*

It seemed to me that the combination of the electrical-voltage field and the reactive metal ions might produce unique cellular effects. We tested the effects of a variety of metallic electrodes on several different types of bacteria in culture. Each metal was tested at voltages ranging from very small to just above the electrolysis level. We found that above the electrolysis level, positive metal electrodes (anodes) killed all bacteria, but they would have also killed any human cells. Only the silver anode killed all bacteria at voltages that would be harmless to human cells.

What we had actually done was rediscover the fact that silver killed bacteria, which had been known for centuries. However, previous clinical applications had consisted of applying very thin metallic silver foil or chemical silver compounds to infected tissues. The silver foil was very inefficient because it gave off only a few reactive silver ions, which did not penetrate the tissues. The silver com-

---

*One can also produce positive metal ions in tissues by the dissociation (or breaking apart) of soluble chemical compounds of a metal. For example, silver nitrate is a compound of silver ions (Ag^+) and the nitrate ion (NO_3^-). When dry, it is a crystalline material ($AgNO_3$); when dissolved, it dissociates into the two original ions, Ag^+ (the positive silver cation) and NO_3^- (the negative nitrate anion.). In this case, *both* ions are capable of reacting with cells. Unfortunately, the NO_3^- forms nitric acid, which is harmful to tissues. The method of electrically generating metallic cations from a positive electrode is the only way to get the metal ion without its accompanying anion.

pounds were damaging to human tissues through the action of the anions. When antibiotics were discovered, clinical uses for silver as an antibiotic were discarded.

Thirty years after the discovery of antibiotics, many bacteria had become resistant to their action, and the clinical characteristics of infected wounds had changed. In my field of orthopedic surgery, local infection of the bone (osteomyelitis) had changed from an infection with one kind of bacteria to an infection with four or more different types of bacteria. This had made the condition much more difficult to treat with antibiotics. It occurred to me that the electrically generated silver-ion technique, which is equally effective against all types of bacteria, might prove useful in such cases.

I began using this treatment in the worst cases of osteomyelitis. These cases involved patients who not only had large open wounds with exposed, infected bone but also had unhealed fractures of the same bone in the infected site. Often, such cases end with amputation, so the new treatment seemed to offer patients their last hope.

There was, however, a complication. In order to stimulate bone healing, negative electricity had to be applied to the nonunion. The scientific data indicated that the application of positive electrical current would not only fail to stimulate bone growth but would dissolve more bone as well. Therefore, before the silver treatment was started, each patient was told that all it could possibly do was kill the infection and let the soft tissues and skin heal. Later, I would use the "right" kind of electricity—a negative electrode inserted into the nonunion site—to make the bone heal.

In each case, the nonunions healed even faster than they would have if negative electrical currents had been used. In addition, the soft tissues and skin healed at the same extraordinary rate. The electrically generated silver ion was doing something more than killing bacteria—it was also causing major growth stimulation of tissues in the wound.

When we finally tracked down exactly what was happening, we found that as human fibroblast cells (which are common throughout the body) were exposed to the electrically generated silver ions, they *dedifferentiated*. They were then able to multiply at a great rate, producing large numbers of primitive, embryonic cells in the wound even in patients over fifty years of age. These "uncommitted" cells were then able to differentiate into whatever cell types were needed to heal the wound. So what we were in fact doing was turning on regeneration in human tissues, which I had thought we would never be able to do.

In our previous studies of regeneration, we had found that in human beings, only bone-marrow cells could dedifferentiate. Because there were so few such cells, we thought that any regeneration in human beings (other than fracture healing) was impossible. The dedifferentiation of the abundant fibroblast cells by electrically generated silver ions may provide us with the means to restore regeneration to human patients.

While I was excited at this serendipitous discovery, my patients could have cared less how the wounds were healed. They were simply delighted at the fact that, at long last, a treatment had worked. Perhaps the most important moral in this story is that it demonstrates that no one can tell in advance what will happen in a scientific experiment.

ELECTROCHEMICAL THERAPY FOR CANCER

Actually, this circuitous pathway led us back to one of our original aims, the control of cancer growth. If the electrically generated silver ion dedifferentiated normal human fibroblast cells, would it also dedifferentiate human cancer cells? If so, we would have a way to duplicate in human beings Dr. S. Meryl Rose's experiments with salamander cancers, in which dedifferentiated cancer cells redifferentiated as normal cells. A lack of funds, unfortunately, prevented us from completing this work. However, we did find that some human cancer cells in culture appeared to dedifferentiate when exposed to these silver ions.

I also had a patient with a severe, chronic bone infection who had an associated cancer in the wound. He refused amputation, which would have been the treatment of choice, and insisted that I treat his infection with the silver technique. After three months, the infection was under control, and the cancer cells in the wound appeared to have changed back to normal. When I last heard from him, eight years after the treatment, he was still fine.

It is important to realize that this is not simply an electrical effect, but the result of the combined action of the electrical voltage and the electrically generated silver ions. It is an *electrochemical* treatment. While we do not have firm evidence at this time, what probably happens is that the silver ion is shaped so as to connect with some receptor group on the surface of the cancer-cell membrane.

After that connection is made, an electrical-charge transfer sends a signal to the nucleus of the cancer cell that activates the primitive-type genes, and the cell dedifferentiates. In that state it awaits instructions as to what it is to become. The process is exactly the same as that in Rose's experiments, except that in this case the dedifferentiation is caused by the unexpected action of the positive silver ions.

This technique obviously requires more study before any clinical use can be made of its antitumor effect. However, it does appear to be a promising lead in an otherwise rather grim picture.

MICROWAVES AND CANCER

Anyone who owns a microwave oven knows that microwaves can cook tissue, be it a steak or a living animal. Because cancerous tissues seem to be slightly more susceptible to heat damage than normal tissues, heat has long been used as a treatment for cancer. Some evidence for this came from misinterpretation of Coley's work (described in chapter 2), assuming that the high fever selectively damaged the cancers. The problem is that of how to heat a cancer deep within the body without heating everything else as well. Microwave technology seemed to provide the answer. With a little engineering, it should be possible to produce narrow microwave beams whose heating effect could be localized. Through the use of two such beams—each slightly below the heat-producing level—that intersected at the tumor site, it would be possible to heat only the tumor.

This idea was vigorously pursued, and problems were uncovered almost immediately. Narrow beams of microwaves were not readily attainable. Also, the heat effect depended on a number of factors, including the amount of bloodflow and the type of tissue exposed. Fatty tissue, for example, absorbed the energy better than other tissues. As a result, both the area and the exact amount of heating were not easily controllable.

Now, some twenty years later, many frequencies and techniques are being utilized. Some even go so far as to involve the surgical implantation of small microwave antennae around the tumor, a procedure that seems rather redundant. If one is already in the body surgically, why not remove the tumor rather than continuing to apply "high technology"?

The technique has not gained wide acceptance in the clinical world, but this is just as well. As we shall see in chapter 8, even at very low power, microwave energy introduced into the body has a number of extremely undesirable effects. Even if we succeeded in "cooking" the tumor, little useful purpose would have been served if the patient then developed other cancers as a result of microwave exposure.

The episode illustrates the blinding effect that high technology can have on the biomedical profession. If a technology is attractive and appears able to do something potentially useful, it is avidly studied and developed. But, too often, no one will think to determine what the exact mechanism of action is or, worse yet, whether the technique has any harmful side effects.

HIGH-VOLTAGE THERAPY FOR SNAKEBITE

This fascinating treatment method comes from the Amazon jungles, where it was developed by unschooled natives who didn't know "it can't be done." The technique is this: as soon as possible after a snakebite has occurred, the spark-plug wire from an outboard motor is disconnected from the plug and held against the bite, while the starter rope is given a few pulls. Anyone who has been unfortunate enough to have his hand on the spark-plug lead while the starter rope is pulled knows that there are a lot of volts in it (about 20,000), but there is very little current. Applied to the snakebite, this voltage not only makes the victim uncomfortable but also seems to inactivate the toxin. The success rate of this low-tech technique is amazingly high, with what would once have been fatal bites now apparently survivable. Since the initial publication of this finding, professional interest seems to have died out.

Exactly what happens to the toxin is unknown. Most such toxins are fairly complex biological molecules. One could postulate that in response to the high-voltage pulse, a molecular rearrangement occurs that inactivates the toxin. However, this has yet to be proved. If this should turn out to be the mechanism involved, it raises an important question about the effect of such fields upon other biological molecules, including those that are normally part of living cellular structures.

A laboratory technique currently in use seems to be somewhat related. This is the electroporation technique, in which a similar short pulse of high voltage is used to cause two adjacent cells to fuse into one. It can be applied when the two cells are of quite different species, producing a kind of cellular "chimera." The useful aspects of the two cells are combined into one cell, albeit a highly abnormal one.

This technique has considerable popularity now. It is used to combine cells that produce useful chemicals, but that don't grow well in culture, with cancer cells that grow extremely well in culture. The end result of the union is a hybrid that grows quickly in culture *and* produces a useful chemical.

NUCLEAR MAGNETIC RESONANCE IMAGING: THE BIOLOGICAL EFFECTS

The use of magnetic resonance imaging to look inside the body (in the same way as done with the CAT scanner) has become a popular diagnostic technique. Hospital administrators often feel that if they don't have this piece of equipment, they are not up to date enough to compete for patients. Because MRI devices are very costly, once they are acquired there is pressure for hospitals to use them to pay back the initial costs. The result is a tendency to order MRI in a wide variety of cases. My observation is that MRI is most often used not to confirm a diagnosis or to look for a specific type of lesion following a thorough physical exam, but rather as a screening test. Frequent justifications for such overusage are that "something may turn up," and that MRI is a harmless procedure with no biological effects.

In reality, in an MRI scan the patient is subjected to a very strong DC (steady) magnetic field, combined with other fields that are oscillating at radio frequencies. While no one has shown specific harm to patients, there are sufficient data in the research literature to indicate that biological effects do occur with these fields. At this time, I do not believe that anyone can give MRI an unequivocally clean bill of health. This does not mean that it should never be used—on the contrary, it is by far the best way to visualize the brain and spinal cord. It can also produce remarkable visualizations of deep-seated tumors in the abdomen.

Again, it comes down to a risk/benefit question. If good medical practice is followed and a tentative diagnosis is made or a reasonable localization of a pathology is obtained, then the use of the MRI technique is well worth the risk. However, the risk involved in its use as an unfocused screening technique does not appear to be justified.

HIGH-ENERGY TECHNIQUES: WHERE WE ARE TODAY

The high-energy techniques appear to be based on the idea that if a little is good, more is better. When one is dealing with subtle, electrical-control systems, this concept is at best invalid, and at worst outright dangerous. Despite its high-tech trappings, modern medicine seems to have progressed little from the days of Galen; simplistic solutions to oversimplified problems are still preferred to the real complexities of life.

This attitude has resulted in a marked increase in the use of high-energy techniques. They are often believed to be truly effective and, at the same time, completely safe. For example, TENS units and their clones are being used in the treatment of headache, toothache, muscle aches and pains due to sports or exercise, postsurgical pain, the pain of heart attacks and angina, and depression, as well as in dental surgery and obstetrical deliveries. The use of electrical muscle stimulators is being promoted as a way to achieve weight loss, body shaping, "cellulite" removal, bust development, wrinkle removal, nonsurgical face-lifts, and muscle development. Recently, the FDA issued a warning about the stimulator devices, stating that no evidence exists for their efficacy in the above conditions; that they should *never* be used on patients with cancer, epilepsy, heart disease, or other serious conditions; and that they might produce miscarriages among pregnant women.

The discoveries of the new scientific revolution have provided not only a basis for a reevaluation of previously discredited medical techniques, but also an opportunity for outright frauds and quacks—as well as well-intentioned but misguided practitioners—to acquire a veneer of respectability. Present scientific knowledge provides no basis for such beliefs as the "ethereal body," the "chakras," the mystical "life force," or the operation of forces that negate the present laws of physics. Any practitioner of energy medicine who postulates that he or she has access to mystical forces or treatments that will "free your body energies" should be avoided.

We have seen how the latest scientific revolution has validated the ancient, preliterate concept of "life energy," not as some mystical, unknowable force but as measurable electromagnetic forces that act within the body as organized control systems. These electromagnetic forces appear capable of being accessed through some of the techniques of the shaman-healers as well as through modern, direct intervention with similar forces. These ideas have led to the development of the new medical paradigm, energy medicine, which is currently being slowly integrated into orthodox scientific medicine.

We have also seen how these control systems relate certain basic functions of living organisms to the electromagnetic environment. In the remainder of this book, we will explore this little-understood but vitally important link. We will see how variations in the Earth's natural electromagnetic field may have played a role in the origin and evolution of life, and how our unwise use of electromagnetic energy has produced environmental changes of unparalleled proportions, with grave consequences for the health and well-being of all living things.

PART THREE
ELECTROMAGNETIC POLLUTION

CHAPTER SEVEN

THE NATURAL
ELECTROMAGNETIC FIELD

More problematic is the claim that evolutionary
change is driven by random *mutations. To place*
pure chance at the center of the awesome edifice of
biology is for many scientists too much to swallow
(even Darwin himself expressed misgivings).
PAUL DAVIES, The Cosmic Blueprint

As noted earlier, preliterate peoples around the world viewed the environment as being filled with mysterious forces that governed their lives. We regard such superstition with amusement, because we "know" there are no such things.

Over the past 500 years, science has given us more and more power to control our lives and destinies. However, researchers now tell us that the complex environment in which we live does, indeed, contain immense, unseen forces—the forces of electromagnetism—that do affect living things. This knowledge has come to light only during the past thirty years, as we have acquired the capability of exploring and conducting experiments in space.

In earlier chapters I referred to the cycles of change in the natural magnetic field of the Earth. In this chapter I will indicate the complexity of these unseen forces and will show how they have played an important role in the origin and evolution of life, and how they still regulate our everyday lives (including our health and well-being).

THE EARTH'S GEOMAGNETIC FIELD

The spinning core of molten iron miles beneath the surface of the Earth creates a dipole magnetic field, much like a bar magnet. However, the energy of the sun distorts and perturbs this simple field into a unique structure, the magneto-sphere.

The sun constantly gives off a solar wind, composed of high-energy atomic particles. These particles travel through space at great speed and crash into the outer layers of the Earth's magnetic field, compressing it until its energy matches that of the solar wind. The area of interaction between these two forces is called the bow-shock region. On the side away from the sun, the magnetic field is drawn out into a long "magnetotail," which stretches away from the Earth far into space.

The Van Allen belts are two areas in this interacting field in which some of the high-energy particles are trapped. These particles constantly spiral between the north and south ends of the magnetic "ducts" in which they are trapped, bouncing back and forth.

In addition to the solar-wind particles, the sun gives off enor-mous amounts of deadly ionizing radiation (such as X-rays) and other high-energy radiation. The magnetosphere shields the Earth from these rays by absorbing or diverting them around the Earth. Without this protection, life could not exist on Earth, as it cannot exist for long in the environment outside of the magnetosphere. Spaceflights beyond its boundaries must be of short duration and must be timed to occur during periods of a quiet solar cycle. As-tronauts outside of the magnetosphere would perish if caught in a solar storm. Thanks to the magnetosphere, we live on a small, protected island in a hostile universe filled with enormous forces.

The Earth rotates on a day-night cycle within this complex field. The magnetosphere does not rotate but remains fixed in space, with one side always facing the sun. Because of this, any given spot on the surface of the Earth is in a constantly changing magnetic field. The daily rise and fall in the strength of this field causes biological rhythms.

This interaction between the energy in the solar wind and the energy in the magnetic field of the Earth results in the generation of enormous electrical currents, with powers of billions of watts. It also results in the production of ionizing radiation and various electromagnetic waves in the extra-low-frequency (ELF) range (between 0 and 100 cycles per second) and in the very-low-fre-quency (VLF) range (between 100 and 1000 cycles per second). This is a "quiet" field, with a steady flow of solar wind from an inactive sun.

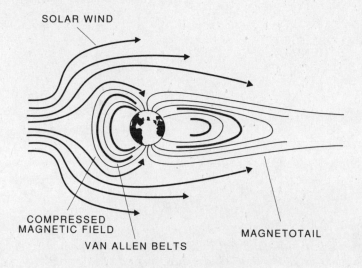

SOLAR WIND

COMPRESSED
MAGNETIC FIELD

VAN ALLEN BELTS

MAGNETOTAIL

FIGURE 7-1. *The magnetosphere is the complex structure of the magnetic field around the Earth, formed by the interaction between the Earth's magnetic field and the solar wind. The lines of the Earth's magnetic field are compressed on the side facing the sun and drawn out into a long "tail" on the opposite side.*

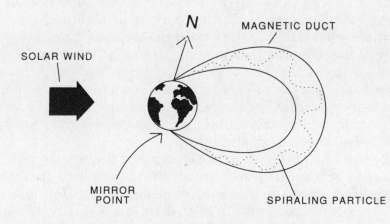

N   MAGNETIC DUCT

SOLAR WIND

MIRROR
POINT

SPIRALING PARTICLE

FIGURE 7-2. *The magnetic duct is formed from adjacent lines of the magnetic field that extend out in space, expanding in size and then returning to Earth at the opposite pole. At each pole the lines of magnetic force come close together, forming a "mirror point." Particles, or electromagnetic signals, trapped in the duct bounce back and forth between the mirror points. This structure protects the Earth's surface from the full force of the sun's radiation. Without it, life on Earth would perish.*

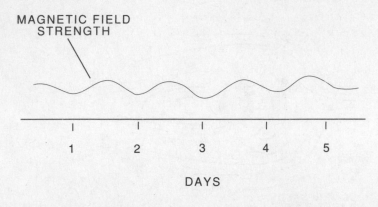

FIGURE 7-3. *Daily rise and fall in the strength of the magnetic field at one spot on the Earth during a quiet period of solar activity.*

MAGNETIC STORMS AND HUMAN BEHAVIOR

The sun is not constant in its activity; rather, it shows a cycle of activity in its energy output that rises and falls on the eleven-year sunspot cycle. During periods of high activity, solar storms are common. These are different from magnetic storms in that they are the result of gigantic eruptions of energy on the surface of the sun (solar flares). Solar flares produce marked increases in the number of energetic particles in the solar wind and in the production of X-rays, streams of protons, and radio waves, which impinge on the magnetosphere and cause great magnetic-field disturbances, called magnetic storms.

During a magnetic storm, the strength of the geomagnetic field fluctuates wildly and increases greatly in strength. This disturbance in the surface magnetic field is often strong enough to cause very high power currents to flow in electric transmission and telephone lines, causing their breakdown. Similar events simultaneously occur in the ionosphere and produce major disturbances in radio and television signals.

Many of us have been fortunate enough to see the northern lights, or aurora borealis. These beautiful, flickering beams and curtains of varicolored lights in the sky are caused by solar storms. The high-energy particles of the solar wind enter the magnetic-field envelope at the polar regions and penetrate down into the upper

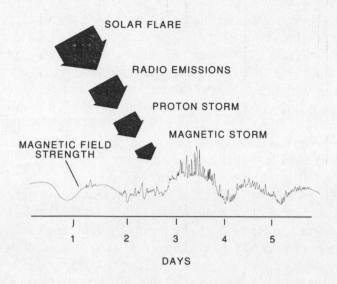

FIGURE 7-4. *The anatomy of a typical magnetic storm. There are many variations, depending on the type of solar disturbance and its duration.*

layers of the atmosphere, where they interact with gas molecules to produce the lights.

If the quiet magnetic field, with its minor daily fluctuations in strength, is sensed by living things, it would appear that magnetic storms should have dramatic biological effects. Long before we had any knowledge of the interaction between living things and magnetic fields, many scientists claimed that such magnetic storms were associated with human and animal behavioral disturbances. These ideas were always dismissed as nonsense, because there was no known physical way that any interaction could take place.

The complexities of the magnetosphere began to be better understood in the early 1960s, primarily as the result of global research done during the International Geophysical Year (1957–58). I had been a volunteer aurora observer during that time and had received the new data on the magnetic-field measurements and relationships. By 1962, my laboratory research had led me to postulate that there really could be some relationship between the geomagnetic field and human biology, particularly in the area of behavior.

In chapter 4, I described how Dr. Howard Friedman and I decided to take another look at the possibility of a relationship between magnetic storms and human behavior. In 1963 we reported that admissions to mental hospitals increased significantly during any week in which a major magnetic storm occurred. Later, we were able to relate the day-to-day behavior of psychiatric patients to lesser changes in the Earth's magnetic field. This raises the possibility that relationships between human beings and the geomagnetic field may not be limited to psychiatric patients; such additional relationships will be discussed later.

Research on the complexities of the magnetosphere is still going on. What I have presented, complex though it may appear, is only a simplified version of the whole picture, which is still emerging. Based on the consideration of the Earth's magnetic environment, physicists have expanded the scope of their research to include the entire universe. Enormous electromagnetic fields and currents are now known to be present in space, interrelated in complex and unusual relationships. Through the work of Hans Alfven, a theory that the formation of the universe resulted from such interactions (rather than from a "big bang") has been gaining ground. This work is of fundamental importance, and where it will lead is anyone's guess at this time.

MAGNETIC REVERSALS AND SPECIES EXTINCTIONS

We now know that the Earth's geomagnetic field has reversed itself many times in the past, with the North and South poles trading places. Humanity has never been through such a "magnetic reversal," but other species have, with dire results. The way in which we learned about this is a fascinating scientific detective story.

The tiny dust particles that enter the oceans, carried by winds or rivers, are deposited on the ocean floor as a sort of "layer cake" of time. This chronological record is studied by dropping hollow tubes into the ocean bottom and removing cores of the sediment. Many of the particles in the cores are magnetite and have tiny magnetic fields. As they are sedimenting through the water, each particle acts like a compass needle and points toward the magnetic North Pole. The alignment of these particles in the cores can be determined by measuring the direction of their magnetic fields, which will indicate the direction of the magnetic North Pole at the time the particles were falling through the water. When this is done, the startling fact emerges that the Earth's magnetic field has often reversed itself in the geologic past. Each reversal is a slow process; in each case, it has taken at least 10,000 years for the new orientation to become established.

The same cores of sea-bottom sediment also contain vast numbers of skeletons of minute sea creatures, the radiolarians. These are single-celled animals that produce hard skeletons around their bodies. The skeletons are complicated structures, different for each species. As these animals die, their skeletons sink to the bottom and become incorporated into the sediment of that era. A chronology of the changes in the species of these animals is thereby established. Examination reveals that from time to time, these animals were subjected to massive species extinctions. These great "die-outs" duplicated the familiar patterns of extinction of other animals, including the dinosaurs. In each such species collapse, the most advanced or evolutionarily developed forms appear to have been the most affected. After each extinction, entirely new types of radiolarians would be established and would grow and flourish, generating many related but more advanced species.

This finding seemed to run counter to our idea of evolution as a steady, upward climb, with new forms of ever-increasing complexity arising from simpler forms. However, in 1967 Dr. Homer Newell of the New York Museum of Natural History reviewed the data on the population size of many different animals and found that there were several periods of time during which many different species had become extinct. These occurred at the end of the Devonian, Permian, Triassic, and Cretaceous geologic periods. It was during the last one that the dinosaurs became extinct.

This description of periods of crises in life has slowly led to a revision of the original Darwinian concept of gradual evolution, replacing it with what we now call "punctuated equilibrium." According to this theory, each period of extinction appears to mark, or produce, a change in the direction of evolution. After the extinction has wiped the slate clean, new forms appear that mark a new pathway of evolution. This raises the question of the causes of past extinctions.

Over the years of study of sea-bottom cores, a strange relationship was seen to emerge. Often, the species extinctions seemed to occur just after a magnetic-field reversal had taken place. Furthermore, if the reversal occurred following an exceptionally long period of a stable field, the species extinction was much more extensive. It appeared as though the life-forms of the quiescent time adapted themselves to that magnetic field; the longer it went on, the greater the impact of the next reversal.

The problem for investigators was how a simple change in the magnetic field, particularly one that took tens of thousands of years to occur, could have any biological effect. At first researchers proposed that while the magnetic poles were changing places, the Earth's field would drop to zero, causing the magnetosphere to collapse and exposing the Earth's surface to the full force of the solar wind and ionizing radiation. The resultant exposure to the full radiation from the sun would kill all organisms. Later, it was found that the field did not actually drop to zero, but merely declined to about half-strength before building back up again. If the magnetic reversals were directly linked to species extinctions, another mechanism had to be operating.

In 1971, a small conference was held on this subject at the Lamont Geophysical Observatory at Columbia University, under the direction of Dr. James Hays. After collecting data on reversals and extinctions, Dr. Hays had found that six of the eight identified extinctions of radiolarians had occurred concurrent with magnetic-field reversals—a relationship considerably above the level of chance. It became evident during the conference that the species extinctions of all animals were somehow linked to the reversals of the magnetic field.

Although the theory that extinctions are the result of impacts on the Earth of comets or asteroids is much in favor at this time, there is clearly an intriguing relationship between the reversals of the Earth's magnetic field and these events. It is, of course, possible that the magnetic reversal was itself the result of an impact upon the Earth, but this seems to be too complex a chain of events to consider. Instead, the question hinges on the idea that a reversal of the magnetic field could alone be a competent cause for subsequent species extinction.

At the Lamont conference, I proposed that reversals could have been accompanied by major changes in the ELF frequencies of the magnetic-field micropulsations, which would have produced behavioral changes that reduced the survival efficiency of the more advanced species. More recently, Dr. Abraham Liboff of Oakland University has proposed that these frequency alterations could have influenced reproduction and produced defective offspring.

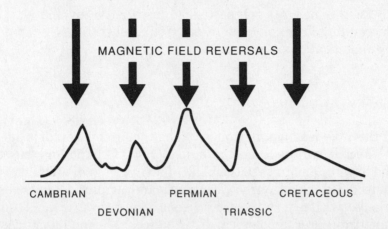

FIGURE 7-5. *The total populations of animal types over geologic time. The time line is not linear, and time between geologic periods is not uniform. The line represents the number of new species when it is rising and the number of species becoming extinct when it is falling. The geologic time periods on the horizontal scale represent the end of each period. Except for the Cambrian, the other periods are those defined by Newell as times of major species extinctions. For example, the peak of the Cretaceous marks the beginning of the decline of the dinosaurs, and the end marks their extinction. The magnetic reversals identified are the major ones—that is, those in which more than 20 percent of the sea-bottom cores showed a reversed pattern. The reversal at the end of the Permian period followed a long period of steady field in which no reversals occurred; the species die-out was exceptionally large. (Adapted from Sander)*

At this time, there is increasing evidence that both of these theories may be valid. They offer one advantage over the cometary-impact theory in that the effects would have been selective, affecting only the more advanced or most sensitive species, and most likely only animals. The cometary-impact theory would require that there was an effect upon all living things, including plants, but the fossil evidence seems to indicate that plants were not significantly involved.

Present evidence from studies on the bioeffects of abnormal electromagnetic fields indicates that such long-term alterations in the frequency spectrum of the micropulsations could have major deleterious effects. Whatever the mechanism, it appears that evolution is *not* simply a random event but is in part driven by changes in the Earth's natural magnetic field, as well as by the abnormal fields resulting from our use of electromagnetic energy for power and communications.

All of this may lead you to wonder when the next reversal is due and what effect it will have on the human race. Our history has been limited to what amounts to the blink of an eye in geological time. We expanded, developed our civilizations, and acquired our scientific knowledge within just the past 25,000 years, in a time of unique freedom from great geological disturbances. We should not assume that this happy situation will continue. The evidence is that we are in the initial stages of a reversal. The average strength of the natural magnetic field has been gradually declining for the past several decades. The micropulsation frequencies are not routinely measured or evaluated, but it appears that no changes have yet taken place. The eleven-year solar cycle is now approaching a peak, and all indices show that this cycle is stronger and more disturbed than any that has previously been measured.

However, the problem may very well be academic. We may have unwittingly produced the equivalent of the greatest reversal ever through our global use of electromagnetic energy. A natural magnetic reversal takes thousands of years to occur. It appears likely that the biological effects of reversals are due to changes in the frequency spectrum of the micropulsations, and we have produced far more than the equivalent change in frequencies in the last fifty years alone. The effects of this change upon human health have already been identified, and they are the subjects of chapters 8 and 11.

THE GEOMAGNETIC FIELD AND THE ORIGIN OF LIFE

Because no one has yet produced life in the laboratory, the origin of life (scientifically referred to as *biogenesis*) is a subject for speculation only. While there are many theories, all require the introduction of energy in some form to accomplish the initial steps in this evolutionary process. Lightning, geothermal energy, and sunlight are favorite candidates for this role.

However, if one speculates on the status of the Earth's magnetic field in the Precambrian era, when life first appeared, some intriguing alternatives may be developed. Basing their ideas on our current knowledge of the atmospheric composition and the status of the geomagnetic field during that time, professors Frank E. Cole of Louisiana State University and Ernest R. Graf of Auburn University have proposed that the magnetic micropulsations would *not* have been "micro," but instead would have been of very great strength. The extra-low frequency of 10 cycles per second would have been particularly strong, perhaps even strong enough to cause the induction of large electrical currents and lightning at the same frequency.

Cole and Graf postulated that this energy could have been used to bring together the initial biological molecules, such as the proteins. It follows that the structures of these chemicals would then have been resonant at 10 cycles per second, and that all living organisms that developed later would have shown a sensitivity to this frequency. This, indeed, appears to be the case among present-day aquatic animals that are electrosensitive, such as sharks, catfish, electric eels, and even that strange mammal, the platypus.

However, this theory does not explain the biggest problem faced by researchers in biogenesis. All complex organic chemicals can exist in two mirror-image structural forms: a right-hand one, and a left-hand one. The amino acids and sugars that make up proteins and other important biochemicals, such as DNA, are of *one type* in all living organisms. We can make amino acids and sugars in test tubes from simpler chemicals, but we always get a fifty-fifty mix of right-hand and left-hand ones. Somehow, when living things originally made their amino acids and sugars as structural units of their bodies, they made only one kind.

This is very important, because we must have only right-hand or only left-hand molecules in order to make a protein or DNA that will work correctly. How living things originally performed this trick is unknown; because the Greek word for hand is *chiros*, the problem is referred to as the chirality problem.

Recently, doctors W. Thiemann and U. Jarzak of the University of Bremen reported that in using a magnetic technique, they were able to generate either all right-hand or all left-hand forms of a number of organic molecules. It is, therefore, possible that the unique magnetic field of Precambrian Earth caused the initial production of only right-hand or left-hand molecules from simple chemicals, making it possible for life to begin.

THE GEOMAGNETIC FIELD AND LIFE TODAY

The ancient idea that there are universal forces governing human affairs does have a basis in reality. Instead of gods and spirits, modern science has substituted electromagnetic fields, but the results are strikingly similar.

The natural geomagnetic field is a complex structure, resulting from the interaction between the simple magnetic field of the Earth and the energy that pours from the sun. It contains vast stores of energy and demonstrates reliable daily fluctuations in strength, as well as longer periodic changes. In addition, it is subject to sudden, massive storms produced by energetic events on the sun. Over geologic time, it exhibits strange reversals of polarity.

All of these energetic changes have biological effects of great importance. Magnetic storms appear to have a direct effect on the operations of the human brain. Magnetic reversals in the geologic past may have been the engine driving the evolutionary process. And the ancient geomagnetic field of the Precambrian era may have played a role in the origin of life.

As we have seen, the geomagnetic field is a shield protecting the Earth from the full force of the sun's energy. Without it, life could not exist. But since humanity has learned how to generate and manipulate electromagnetic forces, we have created other forces beneath this shield, the likes of which have never before existed. Chapters 8 and 11 will deal with the bioeffects of these unnatural electromagnetic fields.

CHAPTER EIGHT

Man-Made Electromagnetic Fields

Of the great construction projects of the last century, none has been more impressive in its technical, economic, and scientific aspects, none has been more influential in its social effects, and none has engaged more thoroughly our constructive instincts and capabilities than the electric power system. A great network of power lines which will forever order the way in which we live is now superimposed on the industrial world.

THOMAS P. HUGHES, *Networks of Power*

O ur modern world began a little over a hundred years ago, when Thomas Edison first demonstrated his electric lamp. By 1882 he had set up the first central generating station, at Pearl Street in New York City, which provided about one-sixth of a square mile of downtown Manhattan with electric power for lights. Edison's system transmitted low-voltage DC current and was limited to short-distance transmission.

At about the same time, Nikola Tesla invented and developed the alternating current (AC) system, which was capable of transmitting far greater amounts of power over much longer distances. By 1894, Tesla generators at Niagara Falls were supplying the city of Buffalo with electric power; four years later, an AC transmission line was operating at 30,000 volts between Santa Ana and Los Angeles, a distance of seventy-five miles. Tesla's AC system is now the one most commonly used around the world. It operates at either 50 or 60 cycles per second (50 Hz or 60 Hz), frequencies that are not present in the normal electromagnetic spectrum of the Earth.

A few years after Edison's electric-lamp demonstration, Heinrich Hertz, a young professor of physics in Germany, demonstrated that an electrical spark jumping across a gap would produce a spark at another, similar gap a few feet away, without being directly connected. While Galvani had observed the same thing a hundred years before, his observation had been totally forgotten. Subsequently, physicists had pronounced that producing "action at a distance" (that is, producing a change at one place by making a change at another place without a direct connection between the two) was impossible. Therefore, before Hertz's time, it was believed that electricity could be transmitted from one place to another *only* by means of a connecting wire.

Hertz was also able to demonstrate that the electromagnetic energy produced by the spark gap was composed of oscillations, or waves, that were part of a much greater electromagnetic spectrum that included visible light. His experiments confirmed the mathematical theory of James Clerk Maxwell, an English physicist, who had predicted exactly such a spectrum of continuously increasing frequencies of electromagnetic energy. In honor of Hertz's contributions, the units of frequency of electromagnetic energy are referred to by his name.

Hertz's apparatus was crude by our standards, but only a few years later, in 1901, it led to Marconi's transmission of the letter *S* across the Atlantic Ocean. Shortly thereafter, the spark gaps were replaced by vacuum tubes, and in 1918 a message was transmitted by "radio" from England to Australia—a distance of 11,000 miles.

By the late 1920s, commercial radio transmissions had become almost commonplace, and electrical power of up to 220,000 volts was being transmitted over distances of hundreds of miles by means of Tesla's AC method. The twentieth century was off to a flying start, and our society would be shaped and molded by the incredible power of electromagnetism.

Today, the economy of every developed nation depends upon its electric-power network. The world is tied together by electronic communication systems that transmit enormous volumes of information practically instantaneously. Such uses of electromagnetic energy are often referred to as our greatest achievement, a triumph of technology that has led to the betterment of society. There is, however, another side to this technology, and that is the subject of this chapter.

OUR ABNORMAL ELECTROMAGNETIC ENVIRONMENT

Our use of energy for power and communications has radically changed the total electromagnetic field of the Earth. Because we cannot directly perceive this with any of our senses, most of us are unaware that it has occurred. Before 1900, the Earth's electromagnetic field was composed simply of the field and its associated micropulsations, visible light, and random discharges of lightning. Today, we swim in a sea of energy that is almost totally man-made.

This change, from the natural electrical and magnetic environment in which life began and evolved to the electromagnetic jungle that now surrounds us, has profound implications for energy medicine. If we sense and derive information from the natural geomagnetic field, it is possible that this unnatural field is producing biological effects that may be harmful. Over the past fifty years, we have more than duplicated the changes in frequency and strength of the micropulsations that may have been associated with past species die-outs, as explained in the last chapter. Our artificial reversal is much greater in extent and has occurred in a much shorter period of time than any naturally occurring reversal. The scientific evidence that will be reviewed in this chapter leads to only one conclusion: *the exposure of living organisms to abnormal electromagnetic fields results in significant abnormalities in physiology and function.*

The growth of electric power and communication systems was slow at first, but since World War II it has been increasing at between 5 and 10 percent per year. In addition, new technologies have appeared. Commercial telephone and television satellite transmitters and relays blanket the Earth from 25,000 miles out in space. Military satellites cruise by every point on Earth once an hour, and from their altitude of only 250 miles, they bounce radar beams off its surface to produce images for later "downloading" over their home countries. New TV and FM stations come on the air weekly. The industry has placed in the hands of the public such gadgets as citizens-band radios and cellular telephones.

Engineers propose gigantic solar-power stations in space,

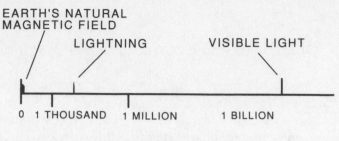

FIGURE 8-1. *The spectrum of the Earth's natural electromagnetic environment. Only the major contributors are listed. There are very small amounts of other frequencies from extraterrestrial sources, and all ionizing radiation higher in frequency than light has been omitted. The horizontal line represents frequency, increasing from zero at left; the numerical listings below the horizontal line refer to cycles per second, or Hertz (Hz). The micropulsation frequencies range from 0 to about 30 Hz. Lightning flashes produce fields in the frequency range of 10–20 kHz (thousands of cycles per second). Visible light is a narrow band in the trillions of Hz. Between each of these normal fields are large blocks of the spectrum that are essentially empty of any electromagnetic frequencies.*

which would relay the electrical energy to Earth by means of enormously powerful microwave beams. Electrical-power transmission lines are operating at millions of volts and thousands of amperes of current. Military services of every country use all parts of the electromagnetic spectrum for communications and surveillance, and the use of electromagnetic energy as an antipersonnel weapon is being studied.

The list goes on and on. We have now almost reached a state in which the entire electromagnetic spectrum has been filled up with man-made frequencies. Our electric-power systems operate at fifty or sixty times per second, just above the highest naturally occurring frequency of 30 Hz. Our microwave beams operate at billions of times per second and are getting ever closer to the trillion-cycle frequencies of visible light. We have filled the previously empty electromagnetic spectrum between these two extremes with man-made radiation that never before existed on Earth. And we did it in less than eighty years.

There are many indices that could be used to show our ever-

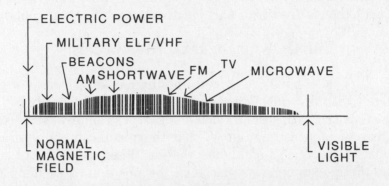

FIGURE 8-2. *Man-made electromagnetic fields of all general types. Some attempt has been made to indicate the number of sources in each area by increased shading, but this should be viewed as approximate only. There is some uncertainty at the very highest frequencies of microwave, because this area is used primarily by the military, and much is classified. It is evident, in comparison with Figure 8-1, that the normally empty portions of the spectrum have been completely filled with large amounts of powerful electromagnetic radiation.*

increasing use of this energy: global electric-power consumption, average transmitting-line voltage, the number of radio transmitters of all types, and so on. All are rising, some at frightening rates. This technological innovation has been considered essential for the advancement of civilization. We can enjoy the advantages of the light switch on the wall and instant access to the news around the world without giving a thought to how these feats are accomplished. The fact is that every power line and electrical appliance produces an electromagnetic field that radiates out from it. And every radio and TV signal that our devices receive has gotten there by means of a similar field given off by the transmitting station. This has resulted in the unseen contamination of our environment with electromagnetic fields of frequencies and powers that never before existed on this planet.

Each electromagnetic field produced by our devices contains energy. The TV pictures on the receivers in our living rooms are produced by the conversion of the energy of the signal given off by the transmitter into electrical energy in the antennas. As we sit watching TV, our bodies are receiving the same energy—as well as the energy from all other TV stations and from FM and AM radio stations, shortwave transmitters, radar devices, electric-power transmission lines, and more. At the present time, there is no place on Earth that is free from this electromagnetic contamination.

THE QUESTION OF HEALTH HAZARDS

The explosive growth in our use of electric power has occurred with few questions being asked about the safety of living things exposed to these abnormal fields. It was simply assumed that the laws of physics guaranteed that there could be no interaction between unseen fields and living things. When questions of safety arose, the questioner was placed in the position of seeming to be irrationally against progress. However, the reason that questions of safety arose was that despite the theories, biological effects were noticed.

In 1928, the General Electric plant in Schenectady, New York, was building an experimental radio transmitter that was to use the highest frequency possible of that time, about 27 MHz. The workers at the plant began to complain of feeling vaguely ill, and Dr. Helen Hosmer of Albany Medical College was called in to investigate. She found that workers' body temperatures increased by as much as two degrees Fahrenheit following an exposure to the transmitter field of only fifteen minutes. In her report, she advised caution in exposing humans to such fields before a complete investigation had been done.

However, the general medical community looked at this phenomenon in a different light. At that time fever was considered to be a "good" reaction to infection and injury. The possibility of artificially inducing fever, or heating localized parts of the body, by this method was attractive. Within two years, radio-wave therapeutic devices (diathermy) were in use and were claimed to be valuable for the treatment of a wide variety of conditions. To the public, then, the technology for producing radio transmissions was not only exciting and useful in everyday life, but could also be used in medical applications. Obviously, it could not be harmful!

The only undesirable side effects noted with radiotherapy were those associated with fever, such as sweating, weakness, nausea, and dizziness. All of these symptoms quickly disappeared when patients were allowed to cool off, and no one looked for any long-term effects.

During the 1930s, the use of this technology for a wide variety of conditions (including injuries, arthritis, migraine headaches, sinusitis, and cancer) increased greatly. Today, diathermy is still used, but for far fewer conditions—not because of the recognition of any harmful effect, but because it has been shown to be of little value for migraine, sinusitis, and cancer.

THE HISTORY OF MICROWAVE FIELDS AND THEIR BIOLOGICAL EFFECTS

One of the great secrets of World War II was the discovery of a method for generating much higher radio frequencies than had ever been possible before. This permitted the development of radar, which played a major role in the ultimate victory of the Allied forces. Early in the use of radar by the U.S. Navy, operating personnel noted the familiar symptoms of body heating, similar to those that had been observed with the 27-MHz radio transmitter at General Electric some years before. Since that effect had been shown to be medically useful as a therapy for various diseases, the heating produced by the much higher-frequency radar was pronounced harmless.

However, since microwave radiation was much more powerful than the lower frequencies in prior use, concerns arose over the exposure of operating personnel because the heating effects could, theoretically, be much greater. In 1942, the navy did the first long-term study, medically examining forty-five radar operators for a year. Nothing more than the usual symptoms from body heating was found. A second study, done in 1945, reached the same conclusion and stated that the heating effect was "of the same magnitude as that used in high-frequency therapy."

In the 1940s, it was already well known that chronic exposure to heat (such as that in operating a blast furnace) produced cataracts, a clouding of the lens of the eye. Therefore, the lens of the eye appeared to be more susceptible to heat than any other part of the body, and the question of cataract formation from chronic exposure to microwave radiation was raised. The initial studies of this possibility were done at Northwestern University immediately following World War II. These studies involved short-term exposure of the eyes of animals to high-power microwaves, followed by an immediate examination of the eye. No pathological effects were found, and it was concluded that exposure to microwaves could not produce cataracts.

However, studies done in that same year by Dr. A. W. Richardson and his colleagues at the State University of Iowa revealed that if the original scientists had waited three days to look at the animals following the microwave exposure, they would have found early signs of cataract formation. Worse yet, when the Iowa group used smaller, non–heat-producing doses of microwave power spread over a few days, the animals developed cataracts forty-two days later. Since the multiple low-power doses of microwave did not produce heating of the lens, it was concluded that microwaves could produce cataracts by means of a nonthermal effect. Even more disturbing, these results indicated that there could be latent effects of microwave exposure, which only became evident much later, after exposures had ceased. While neither of these observations could be explained physically, biologically they did occur.

These reports were only taken to indicate the need for "further research," and between 1950 and the mid-1970s, an impressive number of studies were done on microwave cataracts. The majority of these studies were funded and controlled by the military and were quite obviously designed to show that microwave cataracts were produced by the heating effect. It was not until the early 1970s that a sufficient number of valid studies were done without military funding. These clearly indicated that chronic exposure to nonthermal levels of microwave caused latent cataracts. Evidently, living organisms did not work in the way that many physicists and engineers insisted they should.

In 1973 two Swedish investigators, doctors E. Aurell and B. Tengroth, reported that a survey of workers in a factory where microwave equipment was tested had shown a significant number of clinical cataracts and, more importantly, evidence for direct damage to the nerve elements of the retina. In 1988, Dr. Robert Birge of Carnegie-Mellon University reported that nonthermal microwave radiation could cause a change in the light-sensitive chemicals in the retina. Birge noted that this change was accompanied by the total absorption of the microwave radiation. It was suggested that Stealth-type aircraft could be made totally invisible to radar by being coated with similar chemicals. Because details of this project are now classified, no further information is available, even though the medical implications are far more serious than the production of cataracts.

Because there was a valid physical reason for cataracts to have been caused by the heating effect of microwave radiation, the majority of research was devoted to this aspect. The possibility that nonthermal microwave exposure could produce the same thing was largely neglected because it lacked a physical theory. However, during this same period (from the late 1940s to the 1960s), other, more significant biological effects of microwave exposure were reported.

In 1953, Dr. John McLaughlin, a medical officer for the Hughes Aircraft Corporation, reported to his employer that he had identified between seventy-five and a hundred cases of unexplained bleeding tendency, as well as a significant excess of leukemias and brain tumors among Hughes workers exposed to low-strength microwaves. The corporation believed that it did not have the resources to investigate these potential problems, and since most of the microwave work was on military contracts, it turned the matter over to the military.

A few years after McLaughlin's report, doctors J. H. Heller and A. A. Teixeira-Pinto of the New England Medical Research Institute reported in the British journal *Nature* that a short exposure to pulsed, 27-MHz radio-frequency fields produced chromosomal abnormalities in the cells of the growing root tips of garlic plants. This was the same frequency that had been allocated to the medical profession for use in radiowave therapy, and the study was ruthlessly criticized. Ten years later, observations of these two researchers were confirmed by Dr. David E. Janes and his colleagues at the FDA. Since then, others have reported finding the same effect at nonthermal levels.

THE FIRST THERMAL-EFFECTS SAFETY STANDARD

These reports of nonthermal effects were ignored despite their obvious importance, and the military continued to dominate the field with the "thermal-effects-only" concept. However, some safety standard still had to be developed for the thermal effects. On the basis of theoretical calculations, it was postulated that microwave exposure at a dose of 100 milliwatts of power to an area of 1 square centimeter of body surface would exceed the ability of the blood circulation to carry away the heat produced, and local tissue heating would occur. In 1957, after applying a safety factor of ten, the military adopted a standard for exposure to microwaves of 10 milliwatts per square centimeter. In 1966, the American National Standards Institute accepted the same standard for recommended civilian occupational exposure.

The 10 mW/cm^2 standard became "graven in stone," and for a long time the establishment resisted any efforts to dispute it. Researchers who reported hazardous bioeffects with microwave exposures at levels below this standard were ignored or ridiculed, and their research funds were withdrawn. Nevertheless, the controversy continued, with many independent scientists insisting that exposure to nonthermal levels of microwaves produced major biological effects. The military responded by referring to studies done by "their" scientists and insisting that biological effects of nonthermal microwave exposure were physically impossible.

THE $5 MILLION EXPERIMENT THAT WENT WRONG

In the early 1980s, the U.S. Air Force School of Aerospace Medicine funded a very large, very expensive study at the University of Washington, under the direction of Dr. Arthur W. Guy. In this study, rats were continuously exposed to high-frequency microwaves of 2.45 gigahertz (with one gigahertz equaling one billion hertz) at approximately 0.5 mW/cm^2, twenty times lower than the "safe" thermal level. The exposures lasted for as long as twenty-five months, and 155 different measures of health and behavior were collected.

This appeared to be a well-designed study that would finally answer the question of whether there were any potential hazards to human beings from chronic exposure to microwave radiation. According to Guy, "The results revealed few differences between the exposed and control rats, and those differences for the most part were either not statistically significant or came and went, suggesting that they may be due to chance."

However, one striking observation was noted: "Primary malignant tumors developed in eighteen of the exposed animals but in only five of the controls." Guy hastened to explain that the incidence of cancers even in the experimental group was actually *lower* than normally expected for the strain of rat used in the experiment. He suggested that no hasty conclusions should be drawn, and that a "consensus among most investigators that the only strong evidence for the hazards of microwaves is found at high levels of exposure" was still valid.

This project was widely reported in the press and discussed at scientific meetings, and it was the subject of a major article in the September 1986 issue of *Scientific American* (from which the above quotes have been drawn). A significant aspect of the experiment was not reported either in that article or in the popular press—but at the scientific meeting at which the results of the study were first reported, it was revealed that all of the animals used, both experimental and control, were *gnotobiotic* (a term meaning germ and virus free). This circumstance alone was responsible for a major part of the $5 million cost of the project.

To produce gnotobiotic animals, the young must be delivered by cesarean section under the strictest possible sterile operating-room conditions (much more stringent than those in use in operating or delivery rooms for people). Following delivery, the animals must be raised and then housed in totally sterile environments for the entire duration of the experiment. This type of environment is akin to the decontamination rooms used to house the astronauts after they returned from the moon, or the "bubbles" within which children born without immune systems are housed.

The use of gnotobiotic animals seems to be not only totally unnecessary, but undesirable as well. Neither we nor the laboratory rat normally live in a sterile world, devoid of bacteria or viruses. On the contrary, we live surrounded by uncountable numbers of organisms. We generally do not get sick unless we are injured and bacteria enter the body through the wound, or unless our immunity is inadequate and we get a communicable disease or infection. An experiment on germ- and virus-free animals has no relevance to the real world.

This point becomes even more apparent when two established facts are considered. First, present evidence shows that at least 20 percent of human cancers are caused by viral infection, and the percentage is considerably higher in animals. Therefore, animals that are maintained in a germ- and virus-free state have an incidence of cancer that is much lower than expected. Second, it is well established that exposure to any abnormal electromagnetic field produces a stress response. If the exposure is prolonged, the stress-response system becomes exhausted, and the competency of the immune system declines to below normal. In such a state, animals and humans are more susceptible to cancer and infectious diseases.

One can only conclude that the experiment at Washington was deliberately designed to sharply reduce the incidence of cancer and infectious diseases in the exposed animals. There can be no other reason for the requirement that the animals be gnotobiotic.

Therefore, if we knew these facts in advance, and we wanted to set up a "scientific" project to expose animals to microwaves for a long time but were required to get negative results, we would have only one choice—to use germ- and virus-free test animals. Being gnotobiotic, both the unexposed control animals and the exposed experimental animals would be protected against the usual dangers of infection and cancer. In Guy's study, the fact that the experimental animals had a lower-than-normal incidence of cancer was totally expected. What was unexpected and highly significant was that even with this protection, the cancer incidence in the animals exposed to microwaves was four times that in the control animals.

The well-designed experiment that should have "proved" that microwaves are safe fell into a trap, and the nature of the trap is revealed by the types of cancer that occurred in the experimental group. These were mainly limited to cancers of the pituitary, thyroid, and adrenal glands; these cancers were accompanied by a significant number of pheochromocytomas, which are benign tumors of the adrenal glands. There were no significant cancers of any of the usual tissues.

The experiment was designed to prevent the results of stress, but the planners forgot that it would *produce* stress. Because stress resistance is mediated chiefly through the three glands just mentioned, we must conclude that the microwave exposure produced an extremely high level of stress—so much so that the resultant prolonged hyperactivity of these glands led to their becoming cancerous. Considering the extreme stress experienced by the exposed animals, if the animals had been normal (rather than gnotobiotic) the entire experimental group would have died of infection or cancer long before the close of the experiment.

Some of the 155 biochemical determinations done by Guy in the course of the experiment confirm this interpretation. Plasma cortisol is one of the chemical substances produced by the adrenal glands under conditions of stress, and it was one of the substances measured in the experiment. At the start, the plasma cortisol was equal in both the control and experimental groups; in the early months of microwave exposure, however, cortisol in the experimental group was elevated above that in the control group, indicating that the experimental animals were reacting to stress. By the latter phase of the experiment, the plasma cortisol of the exposed animals was depressed below that of the controls, indicating that the stress-response systems of the experimental animals had become exhausted. This result is exactly as expected for a condition of chronic stress.

These data, which are buried in the multivolume official Air

Force report of the project, were first published in the July-August 1984 issue of *Microwave News*. The experiment was planned cleverly, but not cleverly enough. It clearly indicated that chronic exposure to microwaves, at levels twenty times below the established safe thermal level, produced profound stress and ultimately exhaustion of the stress-response system. Because the experiment involved gnotobiotic animals, this resulted only in a significant increase in cancers of the stress-response glands. Had the experiment been performed under real-world conditions, the results would have been catastrophic for the exposed group of animals.

The Second Thermal-Effects Safety Standard

In 1982, while the Guy study was going on, the American National Standards Institute (ANSI) reviewed the original 10 mW/cm² standard and revised it in light of "new" knowledge. This time, great attention was paid to the theoretical relationship between the wavelength of a radio signal and the length of the human body. It was well known that if the length of the receiving radio antenna was made equal to the wavelength of the desired radio-frequency signal, maximum energy would be obtained from the received signal.

If we say that the average human body is about six feet in length, then frequency range of 80 to 100 million hertz (80–100 MHz) would overlap both above and below six feet. These should be maximally absorbed by the human body, resulting in greater-than-expected heating effects. This frequency range just happens to overlap the 88–108-MHz frequency band used for commercial FM radio. Therefore, with nothing more than a theory (no actual experiments were done to prove or disprove this concept), ANSI adopted a new standard that was frequency dependent but was *still based on the thermal-effects concept alone.*

As a result, the recommended safe levels for exposure to FM were reduced sharply, the microwave-exposure levels were slightly reduced, and exposure levels for frequencies below the FM band were raised significantly on the basis that the longer the wavelength, the lower the possibility of any biological effect. While Guy's data disproved the idea that only thermal effects were possible, that concept still prevails at this time, and all levels recommended as safe are based on the theoretical production of heat in the exposed body. One can only conclude that the ANSI standards are not based upon scientific data and are, therefore, invalid.

MICROWAVES AND GENETIC EFFECTS

While Heller's and Teixeira-Pinto's garlic-root tips seemed to be a long way away from the human being, mice are uncomfortably closer. In 1983, doctors E. Manikowska-Czerska, P. Czerska, and W. Leach of FDA's Center for Devices and Radiological Health reported the effect on the reproductive cells of male mice of exposure to microwaves. They found that sperm production decreased with a short exposure (thirty minutes per day for two weeks) to a nonthermal level of microwaves, and that this was accompanied by significant abnormal changes in the structure of the chromosomes of the sperm. In addition, when the exposed male mice were mated with unexposed females, a significant increase in fetal loss was found. The researchers concluded that chromosomal abnormalities were produced by the microwave exposure at dose rates far below those producing a heating effect. Furthermore, they noted that the mechanism appeared to be a direct effect of the microwaves on the chromosomes themselves.

The most overt expression of genetic defects in human beings is the birth of a child with developmental defects. It is popularly believed that this can only occur if someone in the family has had a similar defect, which has been inherited by the child. However, this is not true. Chromosomal changes in the germ cells or the fetus can be produced by external causes.

In a recent study, doctors Kathryn Nelson and Lewis Holmes of Boston's Brigham and Women's Hospital surveyed 69,277 newborn infants and identified 48 with major developmental malformations. Of these, 16 had no family history of such problems, and the malformations thus appeared to be the result of spontaneous mutations. Since the infants surveyed were born during the years 1972–1975 and 1979–1985, it appears at this time that at least 30 percent of genetic developmental defects in human infants are the result of some external cause. Ionizing radiation (X-ray, for example) is one such cause. The work of Heller and Manikowska-Czerska et al. suggests that abnormal electromagnetic radiation may have the same effect.

In that light, the reports of a relationship between Down's syndrome (a specific chromosomal abnormality) and microwaves are interesting. In 1965, Dr. A. T. Sigler reported in the *Bulletin of the Johns Hopkins Hospital* that children born to fathers who were military radar operators had a significantly higher incidence of this disease. Twelve years later, Dr. B. H. Cohen, also of Johns Hopkins, reported that further study did not confirm these findings but concluded that the link between microwaves and Down's syndrome could not be ruled out.

Over the past few years, Vernon, New Jersey, a little town of about 25,000 in the northern part of the state, has become front-page news. On the basis of the numbers of microwave transmitters, Vernon ranks fifth in the nation, behind New York, Chicago, Dallas, and San Francisco. The incidence of Down's syndrome cases in Vernon is nearly *1000 percent* above the national average. Investigations have been done by the Environmental Protection Agency, the Centers for Disease Control, and the New Jersey Department of Health on the possible link between the excess microwave exposure and the excess incidence of Down's syndrome (and other birth defects), but nothing has been found. The citizen's group that initially raised the issue charges that the investigations have been botched and that the issue has been politicized. My review of the investigation reports supports the townspeople's contention (see the Resources section for this chapter).

MICROWAVES AND BRAIN TUMORS

In 1985, Dr. Ruey Lin of the Maryland Department of Health reported on an epidemiological study of people whose occupations would expose them to higher levels of electromagnetic radiation than would be experienced by the general public. He found that a significantly greater number of the exposed group developed cancer of the brain. Lin also reviewed a study, conducted by the U.S. Navy following the Korean War, that evaluated a possible link between brain tumors and microwave exposure. The navy had compared the rate of brain tumors among people who were radar operators with the rate among members of another group of navy personnel who were not exposed to radar. The researchers had found no difference and had concluded that there was no relationship between radar exposure and brain tumors. Lin's review of this study revealed that the control group in the navy study had actually been exposed to radar to the same extent as the experimental group. Therefore, the navy's conclusion was based upon biased data. Lin recalculated the navy data, using an appropriate control group, and found that there had in fact been a significant increase in brain tumors in the exposed group of personnel.

Shortly after Lin's report was published, doctors Margaret

Spitz and Christine Cole of the M. D. Anderson Hospital in Houston, Texas, reported that "children of fathers employed in occupations with electromagnetic-field exposure were at significantly increased risk" of developing brain cancer before the age of two. This was a particularly chilling report because the children themselves were never exposed, either in the uterus or following birth. The only way they could have had a higher-than-expected incidence of brain tumors was if their fathers' genes had been altered by the microwave exposure and had been passed on to the children, in the same fashion as that of Manikowska-Czerska's male mice.

During the years between 1940 and 1977, there was an unprecedented increase in the use of microwaves. During that same period, the incidence of primary brain tumors among whites rose from 3.80 to 5.80 per 100,000, and the incidence among blacks rose from 2.15 to 3.85 per 100,000. While these data do not prove a direct connection, when taken along with the reports of Lin, Spitz, and many others, they raise valid questions.

It is not possible here to list the many other studies that have lent support to the causal association between microwave exposure and cancers of all types (not just brain tumors) and genetic abnormalities. The scientific data at this time indicate that microwaves have major biological effects at power levels far below those required to cause heating. The majority of these effects are productive of various disease states, primarily cancer and genetic defects, in those exposed and in their unexposed offspring. These diseases are not strange new types unique to microwave exposure; they are instead our old, familiar enemies. The hazard comes from the fact that exposure to microwaves, like exposure to any abnormal electromagnetic field, produces stress, a decline in immune-system competency, and changes in the genetic apparatus. Thus, the levels of exposure that the government says are "safe" are in fact not safe at all.

EXTREMELY LOW-FREQUENCY (ELF) RADIATION FROM ELECTRICAL-POWER LINES

While portions of the American population are exposed to some level of microwave radiation, we are *all* exposed to the electrical-power frequency of 60 cycles from the fields given off by the extensive network of transmission lines and the electrical wiring in our homes and offices. The 60-Hz electric-power frequency lies within the band termed "extremely low-frequency," or ELF, which covers the region of the electromagnetic spectrum from zero (or DC) to 100 Hz.

For very good scientific reasons, it was formerly considered totally impossible for an ELF field to produce any biological effect. First of all, the wavelengths are absurdly long to produce resonance with any living thing. For example, the wavelength of 60-Hz radiation from power lines and appliances is about 3,000 miles. If you apply the antenna-resonance idea to this region, as ANSI did for the FM-broadcast region of the spectrum, the only living organism that could possibly be affected would be a 3,000-mile-long earthworm! Second, because the power in any electromagnetic field is roughly directly proportional to the frequency, it follows that the 60-Hz fields of power lines have an extremely low energy content. As a result, the fields from the electrical-power system were declared absolutely and positively safe.

However, these ELF fields have some interesting properties. They may be transmitted great distances in the cavity between the surface of the Earth and the lower levels of the ionosphere, and they penetrate into the ground and the oceans with ease. These properties came to the attention of the U.S. Navy in the mid-1960s, when the nuclear-missile submarine fleet was being expanded. There was a need for a way to communicate with these vessels around the world without their having to come to the surface and reveal their positions. Because of the unique transmission properties of ELF fields, the navy decided to try using them for this purpose. A very large antenna system, with the code name SANGUINE, was constructed at Clam Lake in rural Wisconsin. It was designed to operate at either 45 or 70 Hz, just above and below the 60-Hz power frequency. Despite its location in the mid-continent, the SANGUINE antenna was able to communicate with submerged nuclear submarines that were as far away as the Indian Ocean. Flushed with this success, the navy proposed to build a truly enormous antenna, which would be buried in the ground under the entire northern halves of Wisconsin and Michigan. This project raised considerable public opposition and political interest, and the navy was required to conduct scientific studies to evaluate the possible biological hazards to crops, livestock, and human beings.

These studies were completed in 1973. The navy recruited a committee of outside experts, of which I was one, to review the results. We met in December of that year at the Naval Medical

Research Institute in Washington, D.C. While a number of results reported to us were positive, one was particularly disturbing. Dr. Dietrich Beischer, working with human volunteers at the Naval Aerospace Medical Research Laboratory in Pensacola, Florida, found that only a one-day exposure to the magnetic-field component of the SANGUINE signal produced a significant increase in the serum-triglyceride levels in nine out of ten subjects. (Serum triglycerides are increased by the stress response and are related to fat and cholesterol metabolism; increases to above-normal levels are a definite cause for concern.)

The navy viewed this result seriously enough to examine the personnel operating the test antenna at Clam Lake. All of the workers showed similar elevations of serum triglycerides. While we could not explain it, it was nevertheless obvious from Beischer's study and the other positive results reported that ELF fields of 45 and 75 Hz had definite biological effects, some of which were potentially hazardous.

The final report of the committee contained a number of recommendations for further study, and the following statement:

> This committee went on record to recommend that the Electromagnetic Radiation Management Advisory Council [ERMAC, the White House agency that had overall advisory capacity in this area] be apprised of the positive findings evaluated by this committee and their possible significance, should they be validated by future studies, to the large population at risk in the United States who are exposed to 60-Hz fields from power lines and other 60-Hz sources.

This statement was unanimously agreed upon by the committee. We were all concerned over the exposure of the civilian population to the 60-Hz power frequency (lying just between the two SANGUINE frequencies), considering the fact that the field strength from the SANGUINE system was a million times *smaller* than that of the field produced by the ultrahigh-voltage electric-power transmission lines. We concluded that large numbers of civilians living in the United States might currently be at risk from these facilities.

After the meeting, the navy denied that it had ever taken place and insisted that it had no knowledge of any scientific studies indicating possible harm to human beings from the operations of the SANGUINE system.

Within a day of returning from Washington, I became aware of the plan to construct ten ultrahigh-voltage power lines in New York State to transmit power from the Canadian James Bay power project into the East Coast power grid. I wrote to New York's Public Service Commission (PSC), the agency with approval authority over the electric utilities in the state, notifying them that the navy had substantial evidence for possible harm to the human population in the vicinity of the proposed lines. I provided them with the name and telephone number of the navy commander in charge of the SAN-GUINE health studies and suggested that they contact him. A few weeks later, I was called by the PSC and told that the navy had refused to talk to them. The result of this was a lengthy series of public hearings on these transmission lines and their possible health hazards.

The final decision by the PSC was to accept my recommenda-tions for a moratorium on constructing these lines and for a five-year scientific study of the possible hazards. This study was to be under the direction of the New York State Health Department; its cost would be $5 million, which would be assessed from the utilities. The utilities fought in court to prevent implementation of this order, but they finally lost, and the program began in 1981.

EARLY POWER-LINE STUDIES

While these events were going on, we started a study in my own lab to look for potential effects of chronic exposure to 60-cycle power fields. We exposed rats continuously to a 60-Hz elec-tric field for three generations, and we determined the infant mor-tality rate and average body weight of the offspring from each generation. We found obvious, significant differences between ex-perimental and control animals in each generation, with the exposed animals having higher infant-mortality rates and lower birth weights than the unexposed controls. These results were identical to those found in rat populations that were continuously subjected to stress.

During this time, I received an interesting letter from British physician Dr. F. Stephen Perry, who worked for the British National Health Service as a family-practice physician in a relatively rural area of England. He had observed that his patients who lived near electric-power lines appeared to have a higher incidence of mental disturbances and suicide. When he mentioned this to various author-ities in Britain, he was not well received. He contacted me for advice

on how to proceed with his own study. My colleagues and I ultimately collaborated with Dr. Perry in an epidemiological study that
showed a significant relationship between power-line field exposure
and suicide in his area. We published the results of the first study
in the scientific literature in 1976, just as the public hearings on the
power lines were beginning. A second study with Dr. Perry, in which
the strength of the field from the lines was measured, was published
as the hearings ended in 1979.

In the meantime, Dr. Nancy Wertheimer, an epidemiologist
at the University of Colorado, was examining the possible relationship between the magnetic field from electrical lines (not the
high-voltage power-transmission lines, but the connecting lines
that are strung on poles along every street). She made a startling
discovery: 60-cycle *magnetic* fields with strengths of only 3 milligauss (three-thousandths of a gauss) were statistically significantly related to the incidence of childhood cancers. This field
strength is many times smaller than the Earth's normal magnetic-
field strength, and it is far below the average strength of 100 milligauss at a distance of approximately fifty feet from the standard
transmission line.

Wertheimer published her data in 1979. Both her paper and our
two papers were immediately subjected to bitter criticism, on the
basis that they simply could not be true; there was no physical link
possible between such extremely weak 60-Hz fields and living organisms.

The New York State Power-Lines Project

When the New York State Department of Health
Power-Lines Project finally got started, it was, to all intents, under
the control of the utilities that were providing the funding. (In present-day "research," this is known as the golden rule—he who has
the gold makes the rules.) When I testified at the public hearings,
I suggested that the one study that should have top priority was a
long-term, large-scale epidemiological study of the New York State
population group living within 200 feet of existing high-voltage
transmission lines. However, rather than doing that study, the Department of Health decided to have Dr. David Savitz of the University of North Carolina repeat the Wertheimer study in the Denver
area. The resources available to Savitz were enormously greater
than those that Wertheimer had had, and I believe that this replica-

tion of her study was undertaken with the expectation of disproving her results.

After five years and the expenditure of almost a half-million dollars, Savitz obtained the same results as Wertheimer. He reported that 20 percent of childhood cancers appeared to be produced by exposure to 3-milligauss power-frequency magnetic fields. The results of the New York State Power-Lines Project were released in 1987. They contained this bombshell, as well as other evidence that power-frequency fields had significant behavioral and central-nervous-system effects, as well as a stimulating effect on cancer-cell growth.

While the final report of the advisory panel to the New York State study was a masterpiece of understatement, the evidence is clear. Exposure to power lines and to other 60-cycle radiation from appliances at levels commonly found in the environment produces an increase in the rate at which human cancer cells grow; an increase in the incidence of childhood cancers; alterations in behavior that are long-lasting, if not permanent; and significant changes in the production of certain vitally important brain chemicals called neurohormones.

The recommendation of the navy's SANGUINE study committee, made in 1973, more than a decade earlier, had finally been validated.

The immediate problem following the release of this report was the need for the Public Service Commission, which had commissioned the study, to act upon the findings of definite health hazards. The 3-milligauss field level was a real embarrassment. The magnetic-field level at the edge of the right-of-way, about fifty feet away from the standard 345-kV transmission line, averages 100 milligauss. These lines constitute the bulk of the transmission facilities in the U.S. If a 3-milligauss safety standard were applied, the right-of-way around almost all transmission lines would have to be considerably enlarged. In addition, many of the distribution lines generate similar strength fields in adjacent residences, and their power would have to be significantly reduced.

The PSC set a "safe" level of 100 milligauss, claiming that the public had accepted this level of risk. This was nonsense. The public had been unaware of any risk before these studies became common knowledge, and even after that the public was told that while some risks might be present, further research was necessary. The public was never asked if it accepted any such risk.

LINKING LOW-FREQUENCY FIELDS AND CANCER

Dr. Wendell Winters of the University of Texas had been contracted by the New York Department of Health to investigate the effects of 60-Hz fields on cells of the immune system. In the course of this work he had exposed human cancer cells in culture to the same fields, without specifically obtaining approval to do so. He reported that cancer cells increased their rate of growth by several hundred percent with only a twenty-four-hour exposure, and that this growth rate was thereafter apparently permanent. The New York State Department of Health sent a team of investigators to Winters's laboratory. They reported that the work was not reproducible and was of questionable validity. The department also funded another investigator to "repeat" Winters's study. This investigator reported that he was unable to duplicate Winters's results; however, he had not done the experiment in the same fashion.

Work was then carried on outside the confines of the New York State study by Winters and his colleague Dr. Jerry Phillips of the Cancer Research and Treatment Center in San Antonio, Texas. Winters's initial observation was confirmed and extended, leading to several recent publications in reputable, peer-reviewed scientific journals. At this time, the scientific evidence is absolutely conclusive: 60-Hz magnetic fields cause human cancer cells to permanently increase their rate of growth by as much as 1600 percent and to develop more malignant characteristics.

These results indicate that power-frequency fields are cancer promoters—that is, they promote the growth of human cancers. Winters and Phillips worked with human cells that were already cancerous, so they could not draw any conclusions as to the possibility that these fields could cause cancer. The promoting effect speeds up the clinical course of any established cancer and makes it that much more difficult to treat.

Cancer promoters, however, have major implications for the incidence of cancer because they increase the number of cases of cancer that become evident. We are constantly exposed to cancer-causing agents in our environment, ranging from carcinogenic chemicals to cosmic rays. As a result, we are always developing small cancers that are recognized by our immune systems and destroyed. Any factor that increases the growth rate of these small cancers gives them an advantage over the immune system, and as a result more people develop clinical cancers that require treatment.

In 1988, Dr. Marjorie Speers of the Department of Preventive Medicine at the University of Texas Medical Branch, Galveston, reported a significant increase in the incidence of brain tumors in workers occupationally exposed to all types of electromagnetic fields. Specifically, she reported that workers exposed to 60-Hz fields in electric-power utilities had an incidence of brain tumors thirteen times greater than that in a comparable unexposed group.

There are many other epidemiological studies indicating a relationship between occupational exposure to electromagnetic fields and cancers of many types. Most of these studies suffer from the shortcoming that the types of fields to which the workers were exposed varied from microwave to the 60-Hz electric-power frequency. It is, therefore, difficult to assign the risk to any particular frequency region; this has been used as an excuse to discount the importance of these studies. In my opinion, this is a specious excuse. The laboratory data clearly indicate a direct relationship between both ELF and microwave fields and cancer. Taken as a whole, the epidemiological data clearly indicate a direct clinical relationship. This view is supported by doctors H. D. Brown and S. K. Chattopadhyay of the Department of Biochemistry at Rutgers University. After surveying the entire literature on the relationship between all electromagnetic fields and cancer, they concluded that "animal carcinogenesis studies and human epidemiological data indicate that exposure to nonionizing radiation can play a role in cancer causation."

LINKING LOW-FREQUENCY FIELDS AND NEUROLOGICAL BEHAVIOR

Another researcher under contract to the New York State Power-Lines Project, Dr. Kurt Salzinger of the Polytechnic University of Brooklyn, exposed rats to 60-Hz fields during fetal development and the first few days of life. The animals were then raised normally until they were ninety days old. At that time they were trained in various learning routines, along with a control group of rats that had not been exposed. Salzinger found that the exposed rats learned more slowly and made more mistakes. He emphasized that the differences were unmistakable and significant, and that they occurred long after the field exposure.

Along the same lines, Dr. Frank Sulzman of the State University of New York investigated the effects of 60-Hz exposure on biological cycles. He found that monkeys exposed to these fields showed a significant decrease in activity levels, judged by the rate at which the animals pressed a lever to obtain food. A surprising finding, however, was that this lower activity level persisted for months after exposure was stopped.

Dr. Jonathan Wolpaw of the New York State Department of Health looked at brain functions under similar circumstances. He measured the levels of neurohormones in the spinal fluid of monkeys that had been exposed to these fields for three weeks. He found that levels of serotonin and dopamine were significantly depressed immediately following the exposure, and that only the dopamine returned to normal levels. Serotonin levels remained well below normal levels for several months.

Both dopamine and serotonin are known to be associated with behavioral and psychological mechanisms. Recently, considerable attention has been directed to the relationship of depressed levels of serotonin to suicide. The data on this research were reviewed and discussed in the *Lancet* (24 October 1987), with researchers concluding that a definite relationship existed. This provides a mechanism for my finding, in conjunction with Dr. Perry, of a direct relationship between power lines and suicide in England.

THE BATTELLE STUDY: THE UTILITY COMPANIES RESPOND

While the New York State power-line study was going on, several other studies were also being done. Of these, the largest was a study conducted by Battelle Pacific Northwest Laboratories, a contract-research laboratory in Richland, Washington. This study was funded by the Electric Power Research Institute (EPRI). In it, minipigs (genetically small pigs) were exposed to specifically constructed duplicates of 60-Hz power-transmission lines for several generations, with a search done for developmental abnormalities (birth defects). The study was actually planned as a duplicate of the small study we had done in my lab on rats exposed to 60-Hz electric fields. The personnel from Battelle visited our laboratory specifically to see our apparatus and discuss our results before they began their own study.

A few months after experimentation had begun, an epidemic occurred in Battelle's pig population. The researchers reported that they had lost almost all of the animals in the exposed group; far fewer in the control group had succumbed. The researchers returned to square one and began the experiment again, but they failed to see the significance of what had occurred. As discussed earlier in regard to Dr. Guy's study on microwave exposure of germ-free animals, the chronic stress produced by field exposure makes animals and human beings much more susceptible to all types of diseases. The first batch of experimental minipigs in the Battelle study were clearly stressed by their exposure to the 60-cycle fields, and so they died in greater numbers than did the unstressed controls.

When the study finally came to a conclusion, the results were clouded with controversy. Battelle claimed that no evidence of harm from exposure had been obtained. However, many critics, including Dr. Richard Phillips, who was the original director of the Battelle study (and who subsequently headed the EPA program on electromagnetic fields), charged that positive findings had been obtained, particularly in the area of developmental defects. Finally, it was revealed that there had, in fact, been an increased incidence of birth defects and fetal malformations in certain generations of exposed animals.

The controversy that erupted resulted in Battelle's duplicating the minipig study using rats. However, despite additional positive findings, the controversy was not resolved. Battelle still insisted that the studies were inconclusive and that further studies were needed.

Recently, Phillips listed the positive findings he believed were obtained from the Battelle studies. These included the following: a marked reduction in the level of nighttime pineal-gland melatonin in rats following exposure to 60-Hz fields for three weeks; a significant reduction in serum testosterone in male rats exposed for three months; changes in the neuromuscular system in animals exposed for thirty days; and an increased incidence of fetal malformations among both rats and minipigs that had been exposed chronically over two generations. These conclusions are at variance with the conclusions of the official Battelle personnel.

MORE DEVELOPMENTAL-DEFECT REPORTS

While the New York State and Battelle studies were under way, Dr. Jose M. R. Delgado established a major laboratory for the study of the behavioral effects of ELF at the Centro Ramon y Cajal in Madrid, Spain. Dr. Delgado is a world-class neurophysiologist, well known for his research on brain behavioral mechanisms and their control by electrical stimulation. This time, however, he was looking for birth or embryonic defects.

Dr. Delgado exposed chick embryos to magnetic ELF fields of three different frequencies—10, 100, and 1000 Hz—and extremely low field strengths. Embryonic malformations were produced with all three frequencies, with most occurring with the 100-Hz fields. At this frequency, major developmental defects were produced by field strengths as low as 1 milligauss. This report caused great consternation and resulted in a flurry of reports from other workers, some supporting Delgado and others reporting no effects whatsoever. The research in Madrid was continued by Dr. Jocelyn Leal, who was able to consistently confirm the original findings.

In 1986, the U.S. Navy entered the picture again. The Office of Naval Research supported an international study, called Project Henhouse, which involved six separate laboratories. All would use identical equipment in an effort to duplicate Delgado's experiment. In June of 1988, the results of this study were reported at a meeting of the Bioelectromagnetics Society. Five of the six laboratories reported that "apparently, very low-level, very low-frequency, pulsed magnetic fields contribute to increased abnormality incidences in early embryonic chicks."

Despite uncertainty as to the mechanisms involved, it is clear that ELF magnetic fields as small as 1 milligauss have the potential to produce developmental abnormalities in growing embryos.

It appears that only two specific functional systems in the organism are primarily influenced by this type of field exposure. These are the brain and the growing tissues of the body, including fetal tissues and cancerous growths. The effects on the brain are mainly functional—for example, behavioral abnormalities, learning disabilities, altered biological cycles, and activation of the stress-response system. In growing tissues, field exposure promotes the growth of cancer cells and increases the incidence of developmental defects in newborns. Strangely, even though there is little cell growth in the brain, cancers of the brain are definitely related to ELF field exposure. There is one system that appears to be possibly related to many, if not most, of these effects, and that is the genetic apparatus. Because genetic alterations had already been shown with microwave exposure, it appeared possible that ELF-field exposure might have a similar effect.

THE GENETIC EFFECTS OF POWER-FREQUENCY FIELDS

A possible explanation for the production of genetic effects by microwaves is that the wavelengths of the microwave radiation may be small enough to have a resonant effect on the DNA molecule or the chromosomes. Orthodox scientists did not believe that this could possibly occur at ELF wavelengths, so they felt that power-frequency fields from high-voltage transmission lines simply could not produce genetic effects. Once again, however, biology has proved that these scientists were missing something.

In 1983, Dr. S. Nordstrom and his colleagues at the University of Umea, Sweden, reported that men who worked in high-voltage electric switchyards fathered a significantly greater number of congenitally malformed children than would be expected. The following year, Dr. I. Nordenson, a colleague of Nordstrom's, examined the chromosomal pattern of the white cells (lymphocytes) in the peripheral blood of similar workers, and found a significant increase in chromosomal abnormalities over normal. The exposure to the electric-power frequency fields (50 Hz in Europe) produced abnormalities in chromosomes of the sperm of the switchyard workers, which resulted in birth defects in their children. This circumstance is identical to that reported by Dr. Manikowska-Czerska for laboratory rats exposed to microwave fields.

Since 1983, Dr. Reba Goodman of Columbia University has been studying ELF-chromosomal effects in both human and insect cells in culture. Her reports indicate a complex picture of many different effects. Some are quite overt, while others have been revealed only through new, sophisticated techniques. In Dr. Goodman's most recent report, presented at the 1988 meeting of the Bioelectromagnetics Society, she indicated that the effects appeared to be different for different frequencies, and that they also differed depending upon the type of cell exposed.

More work needs to be done to sort out the effects and the responsible physical factors. However, there is no doubt that both microwave and ELF frequencies have the capability to influence genetic material during the process of cell division. There is even some indication that DC fields can influence mitosis and chromosomal patterns. This evidence is presented in a later chapter, along with the latest theory for the physical link between ELF electromagnetic fields and the genetic apparatus.

THE SPECTRUM BETWEEN MICROWAVE AND ELF

All of the above reports deal with electromagnetic fields at opposite ends of the nonionizing portion of the electromagnetic spectrum. The power-frequency fields oscillate at less than 100 times per second, while microwaves oscillate at hundreds of millions times per second or more. Nevertheless, it is apparent that the types of biological effects associated with each are virtually identical. It is believed that this is the result of the modulation of the high-frequency microwave signal at lower frequencies.

For example, Dr. W. Ross Adey of California's Loma Linda Medical Center has reported that the release of calcium ions from nerve cells following exposure to 16 Hz may also be produced by exposure of nerve cells to a microwave field modulated at 16 Hz. The microwave alone (unmodulated) has no such effect. The two types of modulation that are biologically important are pulsed and amplitude (see Figure 8-3).

Modulation is the secret of transmitting information by means of electromagnetic fields. AM radio, for example, is amplitude modulated: the radio receiver "demodulates" the signal, removing the "carrier" radio-frequency wave and saving the slowly rising and falling modulation, which is what we then hear as music or voice. Transmission of the carrier wave alone would produce no sound, or a steady tone, depending upon the type of AM radio used. It appears that the body also demodulates the signal when exposed to modulated radio-frequency or microwave fields; the biological effect is that of the low-frequency modulation.

In this view, *all biological effects are produced by ELF frequencies*. This makes good sense, because the body systems that pick up the electromagnetic field are "tuned" to the natural frequencies between zero and 30 Hz. These systems will sense abnormal fields that are close to the normal (between 35 and 500 Hz). The systems then produce an abnormal effect. Microwave radar pulsed at 60 Hz would have the same bioeffect as a 60-Hz field alone, which explains the identical effects seen at ELF and microwave frequencies. It also indicates that all intervening frequencies (VLF, AM

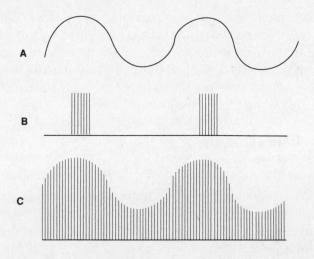

FIGURE 8-3. *Examples of modulation. A is a primary 16-Hz signal, oscillating at 16 times per second. B is pulse modulation (a microwave or radio-frequency signal, pulsed at 16 Hz). The microwave is turned on every 1/16 second and is off the rest of the time. C is amplitude modulation. This is the same microwave or radio-frequency signal oscillating continuously, but with the signal amplitude, or power, rising and falling smoothly at a 16-Hz period.*

radio, FM radio, and TV) will have the same biological effects, since they, too, are modulated.

This intervening portion of the spectrum has received only spotty attention, and the number of reports dealing with specific frequency bands in the radio frequencies is small. Most attention has been directed to the most common sources of exposure, such as commercial AM and FM radio.

In the early 1970s, Dr. William Morton of the Oregon Health Sciences University was asked by the EPA to look into an excessive incidence of uterine adenocarcinoma among residents of a Portland neighborhood that contained an unusual concentration of broadcast towers. The project was expanded to study the relationship between EPA measurements of FM radio fields in Portland and the incidence of several types of cancer in the same area. While no relationship with adenocarcinoma was found, a small but significant relationship *was* found between field intensity in the FM band and the incidence of nonlymphatic leukemia. The EPA took no action on this report.

In 1986, doctors B. S. Anderson and A. Henderson of the Hawaii Department of Health surveyed the city of Honolulu according to census-tract areas. They found that in eight out of nine census tracts containing broadcast towers, the incidence of cancers of all types was significantly higher than in adjacent census tracts that did not have broadcast towers. No action has been taken by the State of Hawaii Department of Health.

THE HAZARDS ARE REAL

While the foregoing review of what has happened in this area since the 1950s may seem complicated, the reports listed are only a small fragment of what is available. In 1963, I was asked to write an article reviewing the scientific literature on the biological effects of magnetic fields. I was able to list a total of forty-four published papers, the oldest from 1892 and the latest from 1962. In the years since then, so many more scientific papers on this subject have been published that I have completely lost track. In 1974, the Office of Naval Research began collecting published scientific reports from around the world in the general area of "biological effects of nonionizing electromagnetic radiation" and publishing abstracts as a digest. This digest is still being published, and the volume of reports has now grown to a point at which more than a thousand scientific papers on this topic are published each year.

The same time period also saw the establishment of three organized scientific societies concerned with this area of inquiry (see the Resources section for this chapter). At present, two of the societies publish international scientific journals, and additional journals are planned. Clearly, this is an expanding area of scientific interest and concern.

The questions with which we began this chapter appear to have been answered: *all* abnormal, man-made electromagnetic fields, regardless of their frequencies, produce the same biological effects. These effects, which deviate from normal functions and are actually or potentially harmful, are the following:

- effects on growing cells, such as increases in the rate of cancer-cell division
- increases in the incidence of certain cancers
- developmental abnormalities in embryos
- alterations in neurochemicals, resulting in behavioral abnormalities such as suicide

- alterations in biological cycles
- stress responses in exposed animals that, if prolonged, lead to declines in immune-system efficiency
- alterations in learning ability

These bioeffects interrelate to influence the clinical state of a human being who is chronically exposed to any abnormal field. For example, the stress effect results in a number of stress-related diseases. If prolonged, this decreases the efficiency of the immune system, resulting in an increased incidence of infectious diseases and cancers. The concurrent effect of the promotion of cancer-cell growth and the increase in the malignant characteristics of these cells leads to an increased incidence of cancers with faster-than-normal rates of growth. (Note that all of these effects are cancer-*promoting*, not cancer-causing.)

However, the effect of abnormal fields upon the genetic apparatus is becoming much more significant. Recent scientific data indicate that many cancers are the result of acquired genetic abnormalities that produce activation of oncogenes, which program cells to become cancerous. Because abnormal electromagnetic fields can produce genetic abnormalities during cell division, it is quite possible that chronic exposure to such fields is a competent cause for the origin of cancers. If this is correct, then the combined promoting and causing effects of field exposure could result in a significant increase in cancers of tissues that are in constant cellular multiplication.

This is compatible with the latest data indicating significant increases in the incidence of specific types of cancers since 1975. According to Dr. Samuel Epstein of the University of Chicago Medical Center, lymphoma, myeloma, and melanoma have increased by 100 percent, breast cancer by 31 percent, testicular cancer by 97 percent, cancer of the pancreas by 20 percent, cancer of the kidney by 142 percent, and colon cancer by 63 percent. All of these cancers are in tissues with continuous rates of cell division.

We are frequently assured that we are winning the war against cancer, and we are told that the death rate from cancer is dropping. This argument is often quoted when the question of the relationship to electromagnetic fields is raised. Some scientists believe that since the total amounts of electromagnetic fields have increased markedly over the past ten years, the overall rate of cancer should have increased at the same rate if a relationship exists between these fields and cancer. This has not happened, so experts claim that there can be no such relationship. While the scientists' statement is true as it stands, what it really means is that the rate from *some* cancers is dropping due to earlier diagnosis and treatment (cancer of the cervix) or to changes in diet (cancer of the stomach), whereas the rates of those listed earlier are increasing. The two effects (decreases in some types of cancer and increases in others) do not quite cancel each other out. The overall incidence of cancer is slowly increasing year by year, and the relationship between field exposure and those types of cancer that are on the increase is evident.

The effect of field exposure on the human fetus is particularly important. The process of development from the original fertilized egg cell to the complex newborn is the result of a carefully timed and controlled series of events involving cell multiplication. Exposure to electromagnetic fields will result in alterations of the rate of growth of these cells and will disturb the delicate balance of normal development. The work done in Madrid by Delgado and Leal on ELF frequencies was completely confirmed by the navy's Project Henhouse and was complemented by the Battelle study on power-frequency fields. Fetal development may be grossly disturbed by field exposure, through the following means: a direct effect of the field upon the speed and timing of fetal-cell division, an effect upon the genetic apparatus of dividing fetal cells, and production of chromosomal abnormalities in the germ cells of the parents.

The distinct possibility that genetic effects of a wide variety may occur with chronic field exposure is disturbing in view of the fact that, generally speaking, such defects are permanent and are passed on to succeeding generations. The apparent inheritance of brain-tumor tendency in the children of men who have been occupationally exposed to these fields may be a significant warning signal. Somewhat more speculative, but still within the bounds of scientific inquiry, is the possibility that the genetic effect of field exposures may have produced alterations in preexisting, disease-producing microorganisms, making them more virulent or resulting in the production of new diseases. That disturbing possibility is the subject of the next chapter.

The interaction of abnormal electromagnetic fields with the brain is now evident, but it remains poorly understood at this time. Some interaction may take place through the pineal gland and the magnetic organ, and there may be direct effects of ELF fields on the nerve cells themselves. These effects are competent to produce behavioral disturbances and learning disabilities. The increase in these clinical conditions over the past decade raises the question of whether our increased use of these fields has played a causal role.

THE ROLE OF ABNORMAL ELECTROMAGNETIC FIELDS

Our ability to generate the entire spectrum of electromagnetic fields is a two-edged sword. With no understanding of the relationship of these fields to living organisms, we have produced a global environmental alteration that has profound implications for human health. Now that we have finally become aware of the serious consequences of this mistake, we have the responsibility to take actions to reduce this threat to future generations. At the same time, as we learn more about the subtleties of the relationship between electromagnetic energy and living organisms, we will acquire increased knowledge of how living things function. This knowledge will better enable us to use electromagnetic energy in our social and economic affairs and in our medical-care systems. The new paradigm of life energy and medicine has the capacity to produce a better world for our children, if we act wisely—and quickly.

CHAPTER NINE

ELF AND THE MIND/BRAIN PROBLEM

I guess, Doctor, your electricity is stronger than my will.

Quote from an experimental patient in *Physical Control of the Mind: Toward a Psychocivilized Society,* by Dr. Jose M. R. Delgado

We believe that our behavior is determined solely by the way our brains integrate information and present it to our "consciousness." We also believe that we have the "free will" to choose either to obey the dictates of our information-processing system or to take another action. In short, we believe that our behavior is *internally* generated by a process of conscious free will. The possibility that behavior is even in part determined by some unperceived external force—one that influences the operations of our brains without our knowledge—has been rejected, primarily on the basis that there is no known external force that could have any such effect.

Even though such terms as *consciousness* and *free will* are quite vague, their meanings are well understood by everyone. We are "conscious" in the sense that we perceive our existence, but exactly how this is done and where it occurs is absolutely unknown. If bacteria lack consciousness, where in the evolutionary sequence does it begin? There is a common tendency to view the brain as a supersophisticated computer. However, no computer has ever been built that displays consciousness (Hal notwithstanding!). Those of us with a tinge of vitalism tend to avoid the mind/brain problem altogether, and we seek refuge in the idea that the brain is an excellent example of living organisms being more than the sum of their parts. Still, we cannot escape the fact that consciousness exists and that it poses one of the greatest challenges to human understanding.

"Free will" is equally difficult to deal with, because it is obviously part of consciousness. We consciously perceive a situation and make a choice as to the course of action we will take. It is this deliberate choosing that we equate with free will. However, our choices are not always rational, and the range of choices available to us is dictated by many external variables, such as culture and society. Nevertheless, we cherish the idea of free will and believe that freedom of choice is the right of all people.

I reject the idea that the brain and its workings, including consciousness, will ultimately be deciphered by computer technology, and I have some doubts that we will ever totally know the exact nature of consciousness in an objective way. However, as I am a scientist, that belief does not preclude my studying the problem. As a starting point, we already know that consciousness resides in, or is associated with, the operations of the brain in the living organism.

If the brain is viewed as a machine, it can be separated into distinct parts, each with specific functions. I doubt that this approach will ever be of use in deciphering the problem, for it appears unlikely that there is a "consciousness center." Consciousness probably resides in the *total* operations of the entire brain and in the systems that integrate the separate parts into a functional unit.

As I noted previously, in the early 1960s I postulated that external magnetic fields could alter the basic operations of the brain by interfering with its normal internal DC electrical current system. I worked in collaboration with Dr. Howard Friedman, a psychologist at New York's Upstate Medical Center, to explore the relationship between the occurrence of magnetic storms and the rate of admissions to psychiatric hospitals. As described earlier, we found them to be significantly related. Admittedly, the subjects of this study were persons with abnormal cognitive patterns, yet the clinical manifestation of the abnormality seemed to be exacerbated by these changes in the geomagnetic environment.

In actual laboratory experiments conducted later, we exposed normal human volunteers to controlled magnetic fields and measured their reaction times. We found that DC magnetic fields as strong as 15 gauss had no effect on reaction time, but that significant and very different effects were produced by exposure to much smaller fields, modulated at 0.1 or 0.2 Hz. These extremely low frequencies (ELF) are present in the micropulsations of the Earth's normal geomagnetic field but are much lower in strength than the main components, which center at about 10 Hz. I concluded that there was some relationship between the ELF in the normal geo-magnetic field (either gross disturbances, such as magnetic storms, or more subtle alterations, such as changes in the relative strength of micropulsation frequencies) and measurable activities of the human brain.

I outlined our theory in a series of presentations and publications in 1962 and 1963. We concluded that

> the magnetic field of the Earth is an important physiologic factor for living organisms. It appears that behavioral changes of an undesirable nature, either quite evident or subtle, may result from exposure to environments having lower or higher field strengths than "normal" or those having either no fluctuation or cyclic fluctuation at frequencies other than those to which we are adjusted.

At that time, the primary scientific endeavor was manned spaceflight, and I cautioned that beyond 250,000 miles out in space, astronauts would be exposed to a magnetic field that was much lower in strength and that lacked the normal fluctuations found at the Earth's surface. I suggested that experiments should be done before such spaceflights to make sure that no subtle psychological alterations would occur that could degrade the astronauts' performance.

As a result of these comments, I was contacted by Dr. James Hamer of Northrop Space Laboratories, who informed me that his group was already involved in this area. He also noted that Dr. Norbert Weiner of MIT, the originator of cybernetics, had been interested in the same subject. Weiner had been involved in a German experiment in which human volunteers were unknowingly exposed to a low-intensity, 10-Hz electrical field. The subjects reported feelings of unease and anxiety when the fields were turned on. Both Hamer and Weiner were working under the assumption that ELF *internal rhythms* in the brain were determinants of behavior, and that pulsing external fields could "drive" these internal rhythms, thereby altering behavior. I corresponded with Dr. Hamer for several years, providing him with the data obtained in my lab; however, I never received any of the data from his experiments.

In 1963 I was contacted by Dr. Dietrich Beischer of the Naval Surface Weapons Laboratory in Silver Springs, Maryland, who was engaged in an extensive project on human volunteers. Beischer was testing the effect on people of long periods of exposure to a zero magnetic field. I visited his lab and found that he was able to completely "null out" the Earth's DC magnetic field, but not the micropulsation frequencies. Therefore, his subjects were not completely cut off from the normal fluctuating components of the natural geomagnetic field. Beischer measured many physiological variables during the project, including those involving psychological factors. When the experiment had been concluded and the data analyzed, Beischer told me that except for some minor psychological variables, no effect had been found.

The experiment could now be redone with both the Earth's steady field and the micropulsations fully nulled out. Weiner's report of the restoration of normal biological cycles under this condition through the introduction of a simple 10-Hz field could then be checked.

However, it appeared that my concept of a relationship between the normal variations in the Earth's geomagnetic field and human behavior was taken seriously by at least certain elements of the scientific community. At the time, their primary interest in this area had to do with possible hazards during space travel. The deeper biological and philosophical implications were not considered.

During the same period of the 1960s, these implications were being taken quite seriously by another scientist. Spanish neurophysiologist Jose M. R. Delgado (whose study of birth defects in chicken embryos exposed to ELF was described earlier) was interested in the physiological basis of behavior and emotions. He confirmed and extended chance observations that had been made a decade earlier and indicated the possible existence of a "pleasure" center in the brain.

ELECTRICAL STIMULATION OF BRAIN AND BEHAVIOR: THE QUESTION OF FREE WILL

Delgado was able to demonstrate unequivocally that minute electrical currents administered through electrodes implanted in the brains of humans and animals produced highly specific emotional and behavioral responses, depending upon the specific part of the brain stimulated. Delgado was a superb showman, and in his most famous demonstration he stopped a charging bull in its tracks by using a radio signal sent from a hand-held transmitter to a receiver implanted in the bull's head. The TV tape of this exhibition was shown around the world.

Delgado discovered the exact parts of the brain in which electrical stimulation produced fear, anxiety, pleasure, euphoria, or rage in human subjects. He found that certain sites produced major personality alterations. For example, stimulation of one such site could cause very proper, reserved young ladies to become flirtatious and sexually aggressive. Other sites inhibited maternal or aggressive behavior. In short, Delgado was able to profoundly alter human behavior through electrical stimulation of discrete areas of the brain.

He promoted the use of this technique for producing desirable behavior in psychiatric patients. When he used the technique on patients who were not too emotionally ill to be unaware of what was happening, they reported that their behavior had changed not because they had wanted it to, but because they had been unable to overcome the power of the electrical signal.

This work ultimately led to experiments in which electrodes were implanted in the pleasure centers of rats' brains. The animals were provided with two levers to press. One of these delivered food, and the other produced electrical stimulation of the pleasure center. Invariably, the rats chose to press the lever that stimulated the pleasure center, *even to the extent that they died of starvation.* The experiments, which were widely reported, were viewed as being "demeaning" to the animals and, by implication, to human beings.

Delgado's work was viewed with open hostility and outrage by the scientific community. These opinions were strengthened by the publication of his book *Physical Control of the Mind: Toward a Psychocivilized Society*, in which he summarized his experiments and his conclusions.

To Delgado, the mind existed *only* in the brain; to postulate its existence as an independent entity was sheer nonsense. He rejected the cherished concept of free will, and he proposed that the mind was a functional entity produced by the electrical operations of the brain. As such, it could—and *should*—be manipulated by external means:

> Because the brain controls the whole body and all mental activities, electrical stimulation of the brain could possibly become a master control of human behavior by means of man-made plans and instruments. . . . To discuss whether human behavior can or should be controlled is naive and misleading. We should discuss what controls are ethical, considering the efficiency and mechanisms of existing procedures and the desirable degree of these and other controls in the future.

Delgado was viewed as promoting human mind control, and his book evoked a storm of protest. Apparently, the mechanistic concept of life works as long as is not extended into such areas as consciousness, mind, and free will.

In the early 1960s I became aware of the work of Dr. H. J. Campbell of the Institute of Psychiatry at De Crespigny Park in London. Campbell was disturbed by Delgado's demonstrations of the power inherent in the pleasure center of the brain, and particularly by its "terrible compulsiveness," as revealed by the rat experiments. He viewed the use of electrodes and internal stimulation as unnatural, and he asked an important question: How do the pleasure areas become activated in normal life?

Through a long series of experiments, Campbell concluded that all new sensory input could stimulate the brain's pleasure centers. His experiments involved providing animals with a way to turn on a stimulus, such as a light or, in the case of aquatic animals, a mild electrical current in the water. He found that initially the animals behaved as though the activity that turned on the stimulus was directly connected to the pleasure center. The aquatic animals, for example, would repeatedly swim through the stimulating apparatus. However, the activity did not continue for long. After the passage of time, it slowed and finally stopped. In other words, the animals became "satisfied" and no longer felt the need for such sensory input. This was quite different from direct electrode stimulation of the pleasure center.

Campbell postulated that as the nervous system evolved and became more complex, there was an accompanying need for the pleasure center to be kept activated by sensory stimuli. He went on to propose that the sensory inputs required for lower animals were simpler and less complex than those for humans. Finally, he suggested that evolution was still progressing, and that humans would slowly lose the need for sensory stimuli to activate the pleasure center. In its place, they would acquire the ability to activate the pleasure center themselves simply by "exercising [the] capacity for thought" and "using [the] sense organs as ancillaries to mental phenomena." Campbell envisioned a utopian society that would be based solely on the pleasures of rational thinking, a view diametrically opposed to Delgado's.

I was interested in Campbell's work because of my own work on the salamander nervous system and its electrical activity. At the time, there was growing interest in so-called electrosensing by aquatic animals, and some researchers were beginning to think that this was a mechanism by which the animals found food in their watery environment. In that light, I wondered whether Campbell's animals had simply been looking for something to eat—would a well-fed salamander swim back and forth through the stimulator?

It also struck me that the animals did not *stay* in the stimulating apparatus, but rather had to swim continuously back and forth through it to obtain "gratification." The fact that the electrical stimulus used by Campbell was a direct current suggested to me that the actual stimulus was ELF pulses of electricity that resulted from the animal's repeat passages through the apparatus.

Unfortunately, my letters to Dr. Campbell were not answered, and I could not locate any of his reports in the scientific literature. While I gave up on these questions, it seemed to me that his basic idea—that normal stimulation of the pleasure centers was via sensory input—was correct. After all, neither human beings nor any other organisms are equipped with electrodes imbedded in the limbic region of the brain. Nevertheless, Campbell's experimental results suggested to me another important conclusion, one that he had missed: *low-frequency, pulsing electrical fields could serve as sensory inputs to living organisms.*

Hamer, Beischer, and I were looking for external fields that could influence behavior, while Delgado and others attacked the problem directly by electrically stimulating the brain. We all learned something. I believe that Delgado, no matter how one views his work, learned the most. Later, he continued with his work in such a fashion as to bring the two lines of investigation together.

Low-Frequency Magnetic-Field Stimulation of Brain and Behavior

Since the mid-1970s, Delgado has been director of the premier Spanish neurophysiological laboratory, Centro Ramon y Cajal. His interest has shifted from direct electrical stimulation of the brain to the broader area of the biological effects of electromagnetic fields. He has studied the influence of specific frequencies of magnetic fields on the behavior and emotions of monkeys, without using any implanted electrodes or radio receivers. While Delgado did not publish any of this work in the scientific journals, its existence leaked out. In 1984 his lab was visited by a friend of mine, Kathleen McAuliffe, then an editor at *Omni* magazine. She was able to observe some of the experiments, and her impressions were subsequently published in the February 1985 issue of *Omni*.

Using very low-strength ELF magnetic fields, Delgado could deliberately produce either sleep or manic behavior in monkeys when the experimental circumstances would normally call for only alertness on their part. In other experiments, he was able to change the effects of actual electrical stimulation of one of the emotional centers of the brain by pretreating the animal's head with another ELF magnetic field. As Delgado noted, the electrical currents produced within the brain by exposure to such fields were hundreds of times lower in intensity than those required to electrically stimulate a nerve cell or those used for direct electrode stimulation. Unfortunately, we do not know whether specific frequencies were always productive of certain behaviors, or whether there were other factors involved. McAuliffe was allowed only a peek behind the curtain; most of the questions I had asked her to put to him remain unanswered.

The mechanistic concept of the brain is essentially that of a "hard-wired" system, but one that could alter its wiring patterns through learning and experience. In this system, information is carried only by the nerve impulse, which is basically the same whether it is transmitting vision, hearing, or information between parts of the brain. The different sensations are the results of signals that are routed from specific organs to specific regions of the brain.

A system such as this, based solely on a single type of signal, could not be perturbed by exposure to ELF fields of the extremely low strengths used by Delgado. The ELF field could change the operations of the pineal gland and possibly the magnetic organ, but the changes would be reflected in alterations in chemical messengers (such as melatonin and serotonin). Behavioral or emotional changes would take considerable time to develop.

It is obvious that ELF fields do not alter the operation of the visual system. For example, standing directly under a 60-Hz high-voltage power line does not produce any visual sensation at all. Except for the gross actions of hearing the hum of the line and possibly noting the static charges developed on the skin, there would be no effect on other system components. On the other hand, expo-

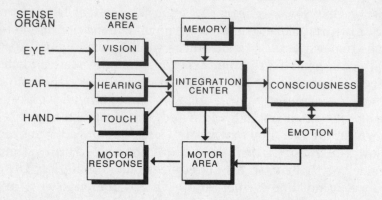

FIGURE 9-1. *Simple block diagram of a "hard-wired" brain. There would actually be many more areas and functions. Emotions would be "wired in" in specific areas, and "consciousness" would be the integrated sum total of all sensory and memory input. The entire system would operate using the same signal (the nerve impulse). The different functions would depend solely on the wiring diagram. (Compare this diagram with Figure 3-4, in which the figure above constitutes only the portion on the lower right.)*

sures to power-frequency fields do produce depression and other symptoms, but only after long exposure and very likely via the pineal/melatonin link.

It is impossible to explain any of the ELF effects on the basis of perception of the field. Likewise, Delgado's observations of prompt alterations in behavior following exposures to low-strength ELF fields cannot be explained by the slow, chemical response of the pineal. The link between ELF electromagnetic fields and the mind must lie deeper—perhaps in the ancient, internal DC system sketched out in modest detail in chapter 2.

LOW-FREQUENCY RADIATION AND ALTERED CONSCIOUSNESS

This DC system contains ELF oscillations that seem to be associated with information flow, and its operations are sensitive to external low frequencies. The overall level of consciousness is regulated by a DC current flow in the primitive, midline structures in the brain. Loss of consciousness can be produced by nulling out or reversing this flow with properly applied external DC currents or strong, steady magnetic fields. There appears to be a roughly linear relationship between the extent of consciousness loss (as judged by the EEG pattern) and the amount of electrical current applied.

In the 1960s, I studied the effect on consciousness of the addition of various ELF frequencies to a baseline DC current. I found that for any current level, a greater loss of consciousness could be produced if I added a very low-strength 1-Hz frequency on top of the DC current. (The 1-Hz frequency alone had no effect on consciousness, and the same strength of DC current produced much less loss of consciousness. Together, the two produced a greater effect.) I later found that any frequency between 1 and 10 Hz had about the same effect. Above 10 Hz, however, the enhancement decreased linearly with the rise in frequency, until at between 20 and 30 Hz the effect was no better than with the DC current alone.

This experiment produced two significant conclusions. First, the DC electrical system in the brain that is related to consciousness is sensitive to very low levels of ELF. Second, *only* the ELF frequency range that makes up the naturally occurring micropulsations is effective.

The concept of a dual nervous system with a primitive DC analog system and a superimposed, sophisticated, digital nerve-impulse system is strengthened by the observation of these ELF effects. The DC analog system *is* influenced by ELF fields. In fact, the interception of the natural ELF fields appears to be one of its functions.

The capacity of ELF fields to alter consciousness and behavior indicates that the interface between the analog and digital systems may be involved in some of the higher nervous functions that we have difficulty explaining with the hard-wired model.

If the digital nervous system—by which we see, hear, smell, taste, feel, and move—is the child of a more primitive system by which we grow, heal, and obey the physical rhythms of our world, then there must be an intersection, a meeting place between the two. Is this the home of the mind, the site of memory, logic, and creativity?

Earlier in this book, we saw how we can consciously cross over this interface and gain access to our own DC systems in order to heal ourselves. By the same mechanism, we may gain control over our thought processes and behavior as well. However, it appears certain that ELF electromagnetic fields, both natural and man-made, can access this same interface and produce significant changes in the total operation of the brain.

The duality of mind and brain has been a philosophical challenge to science for centuries. The dual nervous system and brain, described in this book, can provide us with a new way to view this age-old challenge.

The implications of this work are considerable. It would seem that we may not be the free agents we like to think we are. Our thoughts and actions are, at least to some extent, determined by electromagnetic fields in the environment that we cannot sense and that we remain unaware of to our peril. The darker side of the problem, the ability to control the human mind, now assumes far greater importance than when Jose Delgado first proposed this possibility in 1969. The political and military implications cannot be, and apparently have not been, ignored.

In the next chapter, we will explore the complex interaction between the inner and outer fields.

CHAPTER TEN

LINKING UP INNER AND OUTER FIELDS: MECHANISMS OF ACTION

Science is a hard taskmaster, and in the light of mounting evidence that suggestions of toxicity are for the most part ultimately confirmed by painstaking scientific inquiry, perhaps it is time to reexamine whether scientific standards of proof of causality—and waiting for the bodies to fall—ought not to give way to more preventive health policies that are satisfied by more realistic conventions and that lead to action sooner.

Editorial, *New England Journal of Medicine* (April 1987)

We have seen how living things have been tied to the Earth's natural magnetic field from the time that life began. We have seen, too, how each individual cell—as well as entire organisms—senses and derives timing information from the natural cycles of the geomagnetic field. And we have also seen how human use of electromagnetism for power and communications has produced an abnormal electromagnetic environment unlike anything that existed before.

While the evidence that these abnormal fields have major biological effects is now overwhelming, classical physicists know of no mechanism whereby either normal or abnormal electromagnetic fields can have any biological effects. This has been the major reason why there has been great reluctance on the part of physicists and engineers to accept the biological and medical data as valid. Yet the data currently available are as conclusive as the data relating cigarette smoking and lung cancer. Our knowledge of the latter link is based solely on epidemiological studies. In the area of nonionizing electromagnetic fields, we have both epidemiological and laboratory studies indicating that abnormal frequencies produce increased health risks. However, no effective action has been taken, ostensibly because no mechanism of action has been identified. The governmental agencies that are responsible have discounted the reports of bioeffects because it is thought that these simply cannot be true; "scientific standards of proof of causality" have been stringently applied.

Part of the problem is that the evaluations of the biological and medical data have been in the hands of engineers whose knowledge of biology appears to be minimal, at best. For purposes of developing a mechanism of action, they viewed the living organism as identical to the dead organism or, worse yet, to a water-filled copper sphere. As a result, neither subtle changes in biological functions nor distinctions between effects at the cellular level (as opposed to the total body level) were ever considered.

In this chapter we will consider a ground-breaking theory that explains how both single cells and whole organisms might gain information from electromagnetic fields. This chapter is more technical than the others; you may consider it optional if you find it too difficult. However, the information presented here is the key to unlocking many remaining mysteries, such as cell division, the healer phenomenon, and extrasensory perception.

IONIZING RADIATION, NONIONIZING RADIATION, AND HEAT

When we first began to make use of electromagnetic energy, it was already known that the entire electromagnetic spectrum could be divided into two parts. Frequencies below those of visible light lacked the power to produce ionization of the body's chemical structures, and those higher in frequency than visible light contained sufficient energy to do so.

Ionization is the disruption of an atom or molecule by strong electromagnetic energy, causing the ejection of electrons from the material. The resultant atom or molecule is electrically unbalanced and has a net electrical charge; in this state it is called "ionized." Such ionized molecules are chemically very active and produce abnormal chemical reactions that are damaging to cells. From a physical point of view, it's easy to understand how ionizing radiation can cause bioeffects. By the same token, however, because nonionizing radiation lacks this ability, it was believed that it had no possible bioeffect except the gross production of shock or heat.

In the early 1900s, our technology was limited to the very low-frequency range of the electromagnetic spectrum (radio, for example, operated at a few thousand hertz). As our technology advanced and we began to produce higher and higher frequencies for radio transmissions, it was accidentally discovered that frequencies of about 27 million cycles (27 MHz) and higher could be generated with sufficient power to produce heating of human tissues. This led to the theory that the energy content of any electromagnetic radiation was proportional to its frequency. The higher the frequency, the greater the energy content, and vice versa. For practical purposes, the dividing line for the production of heat by this radiation was set at 27 MHz. Therefore, frequencies above 27 MHz would produce proportionally more and more heating effects, while those below 27 MHz had less and less ability to do so. This was taken to indicate that any biological effect that was not the result of ionization *had* to be the result of heat.

THE "ANTENNA" THEORY AND HEAT EFFECTS

The antenna theory, the first attempt to explain nonthermal effects, was mentioned briefly earlier in this book. Actually, this theory still explained nonthermal effects as heating effects, by postulating the maximal transfer of energy from the radiation to the human body by frequencies whose wavelengths equaled the dimensions of the body. In that way, a field strength below the thermal level was supposed to cause body heating.

There are, of course, problems with this idea (for one thing, we don't spend all our time lying down!). Nevertheless, the EPA became very interested in this concept because the 100-MHz frequency is centered within the 88–108-MHz band assigned to commercial FM stations, which are common in the environment. However, it was soon pointed out that unless a person was very close to the FM transmitter and received adequate power from the signal, tissue heating could not occur even if a resonance situation was produced.

Heat is molecular motion; the faster the motion, the more heat there is. The molecules of every material object are in constant motion, with the rapidity dependent upon the ambient (environmental) temperature. This motion stops only at absolute zero, far below

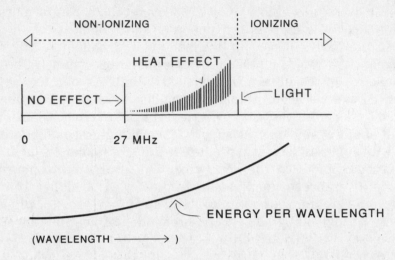

FIGURE 10-1. *The simple electromagnetic spectrum, with its energy relationships and theoretical biological significance. Above 27 MHz, heating would occur, and it would rise with increasing frequency. Below 27 MHz there could be no heating and, consequently, no biological effects. Visible light further divided the spectrum into ionizing and nonionizing portions. The frequencies above light were capable of producing ionization, while frequencies below light lacked that capacity.*

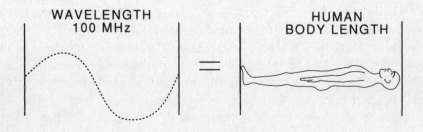

FIGURE 10-2. *Simple resonance between a 100-MHz electromagnetic wave and a six-foot human.*

zero Fahrenheit. The molecules of the body are, therefore, continuously in motion proportional to the body temperature. This is called "kT," or kinetic temperature. In order for heat to have any biological effect, enough energy has to be transferred to make the molecules move even faster. The energy in the environmental FM radiation was simply not enough to produce perceptibly more molecular motion than that of kT, and it therefore could not increase body temperature at all, even if the person was lying down.

According to this concept, microwaves produce their heating effect by causing oscillations of body molecules that are of a size to be resonant with the wavelength of the applied microwave radiation. Unfortunately, this is not the case. Microwave heating is actually produced by an increase in the motion of water molecules. This poses a real problem for the theorists. The size of the water molecule is far too small to be resonant with microwaves. In fact, water is optimally heated by infrared radiation, which has much shorter wavelengths than microwaves. At this time, we really do not know exactly how microwaves produce their heating effect, even in the ubiquitous microwave ovens. A number of theories have been proposed, and the interested reader is referred to a review of this problem by Dr. Jearl Walker in a recent issue of *Scientific American.*

Biological Effects of Radiation without Heat

The problem became much more acute when the results of the navy's SANGUINE studies (see chapter 8) became known. Because the energy in electromagnetic radiation is proportional to the frequency, there was absolutely no way that 45, 60, or 75 Hz could cause any heat effect. Yet, the SANGUINE studies indicated important functional changes in humans. Later, other studies were done at 60 Hz, and additional biological effects were found. By this time the engineers were certain that the biologists were either very poor scientists or certifiable loons! However, the physicists were not so sure.

The work of doctors Susan Bawin and W. Ross Adey of Loma Linda University was the benchmark in this situation. In 1976, they reported that irradiation of living nerve cells in culture with 16-Hz fields produced a measurable increase in the number of calcium ions (Ca^{++}) coming out of the cells. When other labs tried to duplicate this result, they confirmed that ELF did, indeed, produce Ca^{++} efflux, but they reported its occurrence at slightly different frequencies. The resulting argument over whether the efflux was due to chance or whether there was a hidden variable that had not been identified generated more heat than light for some time.

ENTER THE RESONANCE CONCEPT

In 1982, Dr. A. H. Jafary-Asl and his colleagues at the University of Salford in England reported that yeast cells displayed both nuclear magnetic resonance and electron paramagnetic resonance, and that these resonances were different depending on whether the cells were alive or dead. They also found that when living yeast cells were exposed to conditions of nuclear magnetic resonance,* they multiplied at twice their normal rate—and the daughter cells were half as large as normal! Perhaps a more complex type of resonance was part of the answer, after all.

The advantage of complex resonances such as nuclear magnetic resonance is that the energy in the field is concentrated upon single physical entities (such as the nuclei of certain atoms), rather than being spread among all the cells of the body.

In 1985, Dr. Carl Blackman of the EPA and Dr. Abraham Liboff of Oakland University, working independently, integrated the reports of Jafary-Asl and the attempts to duplicate Bawin and Adey's experiments. They concluded that the strength of the local steady-state magnetic field of the Earth at the site of each of the laboratories was the hidden variable that determined the different frequencies reported.

Both Blackman and Liboff suggested that the mechanism involved was a specific type of resonance, cyclotron resonance (which has nothing to do with the cyclotron, an early type of particle accelerator used in atomic physics). When they applied the mathematical equations for cyclotron resonance to the different frequencies reported by the different laboratories, along with the respective strengths of the local magnetic fields, they found the same result.

---

*Nuclear magnetic resonance is now a familiar diagnostic tool in medicine known as magnetic resonance imaging, or MRI (see chapter 6). It is one of a group of complex resonance phenomena based upon the relationship between a constant, magnetic field and a time-varying, or oscillating, electric or electromagnetic field affecting the motion of charged particles, such as ions, charged molecules, atomic nuclei, or electrons in the body.

The Ca^{++} efflux was the result of cyclotron resonance between the frequency of the applied electric field and the strength of the Earth's local magnetic field at each separate laboratory.

Cyclotron resonance can be explained as follows, albeit in a somewhat simplistic fashion: If a charged particle or ion is exposed to a steady magnetic field in space, it will begin to go into a circular, or orbital, motion at right angles to the applied magnetic field. The speed with which it orbits will be determined by the ratio between the charge and the mass of the particle and by the strength of the magnetic field.

We know the frequency of rotation (the number of times per second that the particle completes a full rotation) from the equation relating the charge/mass ratio of the particle and the strength of the magnetic field. If an electric field is added that oscillates at exactly this frequency at right angles to the magnetic field, energy is transferred from the electric field to the charged particle.

If the direction of the electric field is slightly off from the right angle, the particle will move in a spiral pathway.

We can substitute an oscillating magnetic field for the electric field and still obtain cyclotron resonance. However, it must be applied parallel to the constant magnetic field.

Cyclotron resonance may be produced any time there is a steady magnetic field combined with an oscillating electric or magnetic field acting on a charged particle. Many of the activities of living cells involve charged particles—such as the common ions of sodium (Na^+), calcium (Ca^{++}), and potassium (K^+)—acting on or passing through the cell membrane. Cyclotron resonance has the ability to transfer energy to these ions and to cause them to move more rapidly. These effects will change the function of living cells by enabling the ions to pass through the cell membranes more effectively or in greater numbers.

Cyclotron resonance is a mechanism of action that enables very low-strength electromagnetic fields, acting in concert with the Earth's geomagnetic field, to produce major biological effects by *concentrating the energy in the applied field upon specific particles*, such as the biologically important ions of sodium, calcium, potassium, and lithium.

The equation for cyclotron resonance says that as the strength of the steady-state magnetic field decreases, the frequency of the oscillating electric or magnetic field needed to produce resonance also decreases. This is particularly significant when the average

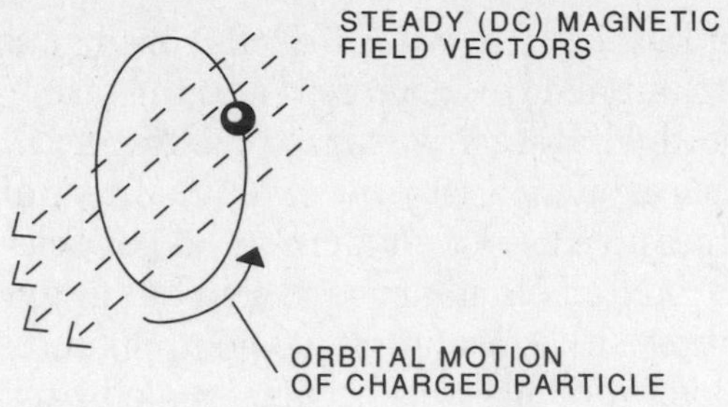

STEADY (DC) MAGNETIC
FIELD VECTORS

ORBITAL MOTION
OF CHARGED PARTICLE

FIGURE 10-3. *The orbital motion of a charged particle at right angles to a steady-state magnetic field. The speed of rotation in orbit is dependent upon the charge-to-mass ratio of the particle and the strength of the magnetic field: the lower the magnetic-field strength, the slower the speed of rotation.*

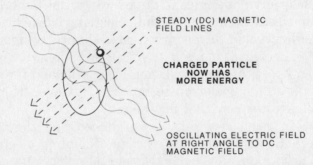

STEADY (DC) MAGNETIC
FIELD LINES

**CHARGED PARTICLE
NOW HAS
MORE ENERGY**

OSCILLATING ELECTRIC FIELD
AT RIGHT ANGLE TO DC
MAGNETIC FIELD

FIGURE 10-4. *Application of an oscillating electric field at right angles to the magnetic field and with a frequency equal to the speed of orbital rotation of the particle will transfer energy to the particle.*

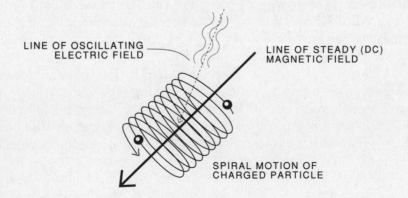

LINE OF OSCILLATING
ELECTRIC FIELD

LINE OF STEADY (DC)
MAGNETIC FIELD

SPIRAL MOTION OF
CHARGED PARTICLE

FIGURE 10-5. *Spiral motion produced by application of the resonant electrical field at less than 90 degrees to the magnetic field.*

strength of the Earth's magnetic field (between 0.2 and 0.6 gauss) is put into the equation: the frequencies for the oscillating fields that are needed to produce resonance with the biologically important ions turn out to be *in the ELF region.*

The ELF frequencies—0–100 Hz—become the most significant part of our electromagnetic environment. The apparent ability of the body to demodulate all higher frequencies, including microwaves, substantiates this. Cyclotron resonance provides an understandable and valid mechanism of action for the biological effects of both normal and abnormal electromagnetic fields.

Doctors John Thomas, John Schrot, and Abraham Liboff, working at the U.S. Naval Medical Research Center (Bethesda, Maryland), first tested this theory using rats that were exposed to a field producing resonance with the lithium ion. They chose lithium because it is naturally present in the brain in very small amounts. It has a calming effect and is used as a medication for the manic phase of manic-depressive psychosis. Thomas and his colleagues predicted

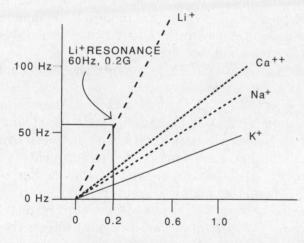

FIGURE 10-6. *The relationship between the strength of the steady magnetic field (horizontal axis) and the frequency of the oscillating electric field (vertical axis) needed to produce cyclotron resonance for several important ions. In the Thomas, Schrot, and Liboff experiment, the 60-Hz oscillating field required a steady-state magnetic field of 0.2 gauss to achieve cyclotron resonance with the lithium ion (Li+). The range of strength of the Earth's natural magnetic field is 0.2 to 0.6 gauss, depending upon location.*

that the cyclotron-resonance effect on the normally present lithium ions would increase their energy level, producing an effect equivalent to a medicinal dose of lithium. The exposed rats should show a depressed behavior as compared to the control rats.

Because the study was supported by the New York State Power-Lines Project, the researchers used an oscillating magnetic field at the power-line frequency of 60 Hz and a controlled magnetic field of 0.2 gauss (the low end of the Earth's average field strength). This combination is resonant with the lithium ion. The rats in the resonant field exhibited much less activity and were more passive and submissive than the nonexposed controls—a result equivalent to the result that would be obtained if the animals were given large doses of lithium.

Since then, more extensive studies have been done, all of which have supported the cyclotron-resonance theory. This theory has been extended in ways too complex to be discussed here. While some criticisms have been raised, these have concerned minor points and do not diminish the great value of the entire concept.

This is not to say that other types of complex resonance, such as nuclear magnetic and electron paramagnetic resonance, do not have equally important biological effects; they probably do. They just have not been so well studied at this time.

The importance of the entire resonance theory cannot be overemphasized. It provides logical mechanisms whereby single cells and specific organs, such as the pineal gland, may intercept and derive information from electromagnetic fields. The theory also appears to be applicable to the basic relationship between living things and the Earth's normal electromagnetic environment.

In 1984, I proposed that resonance between the Earth's natural steady-state field and the micropulsation frequencies might provide the timing mechanism for cell division. Since the resonance theory is based on frequency and not power, it permits effects from vanishingly small fields, such as that observed in 1978 by Dr. Yu Achkasova at the Crimean Medical Institute in the USSR. The sun's magnetic field is organized into sectors, much like the segments of an orange. Alternate sectors have their fields directed inward and outward, so that as the sun rotates a slight change is produced as each sector boundary crosses the line connecting the sun with the Earth. Achkasova observed a rhythm in the multiplication rate of bacteria in culture that coincided with the passage of each solar-sector boundary—an incredibly minute alteration of the field at the surface of the Earth.

Biological Resonance: How It Works

The electromagnetic-resonance concept may provide an intriguing link to a number of little-understood and disputed phenomena, such as extrasensory perception and the ability of "healers" to diagnose and treat patients. In both of these activities, the participants may be unconsciously using an innate biological mechanism similar to that of magnetic-resonance imaging.

MRI depends upon the same principles as cyclotron resonance. The latter requires both a steady-state magnetic field and an oscillating electrical field acting on a charged molecule or atom. Nuclear magnetic resonance is quite similar, except that the nucleus of the atom is resonated. The magnetic imagers in use today resonate the nuclei of hydrogen atoms in the body. The two fields combined transfer energy to the hydrogen nuclei. The image is formed by producing the resonance, exciting the hydrogen nucleus, and then stopping the oscillating field. The nucleus immediately returns to its unexcited state, giving back the energy as an electromagnetic signal that is sensed by the imager.

In theory, all types of complex resonance may be used in the same way. The resonant state is produced when the right combination of fields transfers energy to a certain component of the body. The fields are then shifted out of resonance, and the excited component releases the energy as an electromagnetic field that is sensed by appropriate devices.

In clinical use of nuclear magnetic imaging, this process is repeated many times. The returned signal is stored in a computer, which gradually builds up a three-dimensional image of the interior of the body. Different organs are seen on the basis of their water content and the state of their hydrogen atoms. Hydrogen nuclei in cancer tissue have a different signal than those in normal tissue, and they return a different signal. Clinical magnetic scanners use very strong magnetic fields and correspondingly high radio-frequency fields as the oscillator. This is done to produce a high-resolution image with great detail. While lower field strengths and lower frequencies would also work in imaging, they would yield little detail.

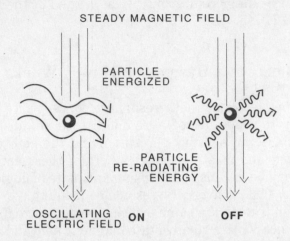

FIGURE 10-7. *When both fields are on, the particles become energized. When the oscillating field is turned off, the particles reradiate the energy as an electromagnetic field that can be sensed.*

MAGNETIC RESONANCE AND THE "HEALER PHENOMENON"

Evidence that healers can diagnose the site of a disease process by giving off some type of electromagnetic field from their hands was presented in chapter 4. I have theorized that the field given off by the healer induces electromagnetic resonance of some body component, and that the healer senses the returned signal. The healer gradually builds up an image in his or her mind that is similar to that of the magnetic imager but has lower resolution. This theory does not require that the hydrogen ion be the target particle. It could be any one of a number of biologically significant ions, or it could be a specific molecule, such as an enzyme or a peptide.

For the healer to be able to diagnose the site of a pathology, the signal that is returned to the healer's hand from pathological tissue must be different from the signal of the same tissue when normal. Such a difference could be produced by a deviation from the normal amount of target ions or particles present or by an abnormal electronic state of the same target ions or particles. It may be possible to explain the therapeutic effect produced by healers on a similar basis. If verified, this could lead to the development of specific, clinically useful devices. We should not only think of influencing specific ions within an area of pathology, but we should also consider some resonant action between the fields given off by the healer and the intrinsic electrical-control systems within the body. While these currents have a vanishingly small flow, they would nevertheless produce local DC magnetic fields within the tissues. These fields, in combination with a frequency field given off by the healer, could result in resonance with other charged particles.

Because this theory rests on complex resonance, the Earth's steady magnetic field must be involved. The oscillating field from the healer should then be in the appropriate ELF range. Because the strength of the Earth's field varies on a diurnal cycle, the resonant frequency range should also vary slightly, and the healer's emitted frequency should shift accordingly. However, it would appear unrealistic to expect a healer to perform with the precision and capability of a frequency generator. Therefore, the range of frequencies possible from the healer is probably limited to that required by the natural variations in the Earth's steady-state field.

If this theory is correct, healers should be able to work with greater precision and display greater diagnostic and healing abilities during times of quiet geomagnetic-field conditions. Conversely, they should be adversely affected by periods of magnetic storms and perhaps by locations in which man-made, high-strength, steady magnetic fields or competing ELF fields are present. None of these possibilities have yet been explored in relation to the healer phenomenon.

MAGNETIC RESONANCE AND EXTRASENSORY PERCEPTION

All of our radio-communication systems rest on resonance phenomena, albeit the simple wavelength type. When we "tune" a radio receiver, we are simply changing the frequency at which the internal circuits resonate, and we receive only that signal in which we are interested. This system is not highly sensitive to magnetic fields except during magnetic storms, which change the characteristics of the ionosphere and either interfere with reception of specific frequencies or enable us to detect signals from much greater distances than normal.

The phenomenon of extrasensory perception (ESP) has been likened to biological radio, with the "sender" giving off an electromagnetic signal that is sensed, in some fashion, by the "receiver." However, application of radio technology to explain this phenomenon has not been successful. It is difficult to explain the distances of transmission when the measurable fields given off from the human brain are almost imperceptible and when the tuned circuits for the transmitter and receiver have not been identified.

However, we now know that ESP of a variety of types is enhanced by a quiet geomagnetic field and adversely affected by a disturbed field. This finding was presented at the session on parapsychology at the Annual Meeting of the American Psychological Society in 1986 by doctors Michael Persinger, Marsha Adams, Erlendur Haraldsson, and Stanley Krippner. Each had worked independently of the others and had used different techniques. Nevertheless, all had arrived at the same conclusion: the phenomenon of extrasensory perception is interfered with by a disturbed geomagnetic field.

This connection could indicate that a complex resonance phenomenon is involved, with the geomagnetic field serving as the steady component. The exquisite sensitivity of the resonance process certainly would assist in dispelling, to some extent, the problem of transmitting and receiving the extremely low power signals that must be involved.

However, the data may also be interpreted in other ways. The phenomenon of ESP requires three components similar to radio-communication systems: a transmitter, a signal that spans distance, and a receiver. If magnetic forces are involved in this process, a disturbed geomagnetic field could interfere with the ability of the transmitter or the receiver to operate or with the propagation of the signal itself across distance.

The apparent ability of the signal to span long distances is a problem if simple radio technology alone is considered. However, ELF signals are known to be transmitted in magnetic ducts over long distances. This transmission is, in fact, associated with an *increase* in signal strength. Magnetic ducts are formed by adjacent lines of the magnetic field extending from the north to the south magnetic poles. The problem is that this can be used to explain only north-south transmissions of ELF signals. We continue to discover new aspects of the Earth's magnetic field, however, and it would appear reasonable to keep these possibilities in mind.

Resonance theory provides us with clues about the types of experiments that could yield some understanding of the mechanisms involved. The most pressing problem of ESP is its lack of reproducibility. On occasion it works with astonishing precision, while at other times it simply cannot be replicated. In science, this is cause for rejection of the entire concept: if something cannot be replicated in a laboratory, it doesn't exist. Knowing the potential relationship of ESP to the status of the Earth's magnetic field might enable us to remove this lack of reproducibility by disclosing the hidden variable. Furthermore, it would permit us to design experiments to explore this relationship in detail.

MAGNETIC FIELDS AND CELL DIVISION

If the cellular events in mitosis are magnetically active and are dependent upon some resonance effect of the Earth's natural (or normal) fields, then the presence of man-made (abnormal) fields could lead to disturbances in mitosis or to genetic abnormalities. The evidence is clear that all cells in the process of active cell division are directly affected by exposure to both ELF and microwaves. Unfortunately, our knowledge of the physics involved in the process of mitosis is minimal, and experiments on the effects of field exposure on cell division have been limited to simply observing changes in the rate at which it proceeds or in the production of abnormal chromosomes.

However, one intriguing observation was made in 1980 by Dr. David Cohen of the National Magnet Laboratory at MIT. Dr. Cohen reported that the SQUID* magnetometer detected a steady magnetic field from human hair follicles. The cells of the hair follicle are constantly in mitosis, but Cohen did not check on the possibility that mitosis was producing the magnetic signal. As a result, several different experiments that could have evaluated this possibility have not been done. Because no one has yet done the simple experiment of directly observing the process of mitosis through a microscope while the cell is exposed to an external magnetic field, at present we can only speculate about what would be seen. We obviously need to pry much more deeply into this important relationship.

If one watches cells through a microscope, the process of cell division appears lengthy, averaging about twenty-four hours. Dur-

*SQUID=superconducting quantum interference device.

ing most of this time, the cell is engaged in doubling its amount of DNA so that there is enough to make two new cells. Dr. Abraham Liboff was the first to discover that this process can be speeded up by exposing the cells to ELF magnetic fields. The actual process of cell division is a complex event that takes only a few minutes to complete. Mitosis (see Figure 10-8) is the final stage of this cycle and is the only stage that can be seen under a microscope. The same process occurs in cells in both the body and in culture.

Using special techniques, it is possible to synchronize this process in all, or most, of the cells in a culture so that they go into mitosis roughly simultaneously. One can also stop the process at the stage at which the chromosomes are just beginning to "pull apart." This technique is used to count and characterize the individual chromosomes and sometimes to reveal abnormalities, as first reported by Heller (see chapter 8).

Dr. Martin Poenie of the University of California, Berkeley, has recently shown that major changes occur in the calcium ions within the cell during the anaphase stage of mitosis. Conceivably, the field exposure may interfere with this process via a cyclotron resonance effect.

It is also possible that some of the complex structures formed during mitosis may have magnetic properties of their own. All substances are magnetic to some extent, because the spin of electrons around the nucleus of any atom is equivalent to a tiny electrical current and will produce a corresponding magnetic field. Magnetic substances may be ferromagnetic (producing a magnetic field on their own), paramagnetic (lining up parallel to the field lines of an external magnetic field), or diamagnetic (turning to a right angle to the field lines in an external magnetic field). These different types of magnetic materials were discovered by Michael Faraday in the late 1800s. Faraday exposed many different substances to nonuniform, or inhomogeneous, magnetic fields, in which the field lines are not parallel to one another. The exact classification depends upon the atomic composition and structure of the materials, and the actual situation is much more complex than that presented here. Nevertheless, this simplistic classification indicates the complexity of the magnetic properties of matter.

If it is somewhat absurd to consider chromosomal effects from ELF fields, it is *totally* absurd to postulate that DC fields could have any such effect. As a result, the majority of scientists in this area have avoided studying DC field effects. However, while searching

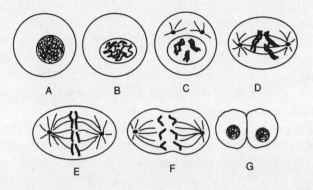

FIGURE 10-8. *Normal sequence of mitosis. A shows the "resting" cell; during this time the DNA is being duplicated. In B, the chromosomes begin to appear. In C, the chromosomes pair up, and the centrosomes appear (for clarity, only three chromosomes are shown). In D, the chromosomes begin to move to the center of the cell. In E, the chromosomes are lined up in pairs along the center line of the cell. In F, it appears as though the strands (called spindle fibers) coming from the centrosomes retract and pull the two pairs of chromosomes apart. In G, two daughter cells are formed, each with exactly one of the original pairs of chromosomes. Scientifically, B, C, and D are called the stage of prophase; E is the stage of metaphase; and F is called anaphase.*

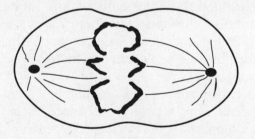

FIGURE 10-9. *Abnormal bridging formed between the pairs of chromosomes during anaphase, following exposure to 27-MHz radiation. It appears that the two sister chromosomes have difficulty in completely separating. In this case, the two daughter cells will not have identical amounts or contents of genetic information, and oncogenes may be exposed.*

the literature in the 1960s I found several earlier reports of DC magnetic-field effects associated with cell division. In 1938, Dr. C. G. Kimball reported that inhomogeneous DC fields of only a few gauss could produce statistically significant decreases in the growth rate of yeast cells. Kimball also reported that homogeneous fields with strengths as high as 11,000 gauss did not change the normal growth rate of these cells. Therefore, the effect had to be due to the inhomogeneous field itself.

In 1978, I experimented with the effects of DC electrical fields (also called electrostatic fields) on growing cancer cells. We implanted cancer cells into mice and then exposed the mice to an inhomogeneous DC field for several weeks. We set up two different conditions: one group of animals was exposed to a field that was horizontal (parallel to the ground), and the other was exposed to a vertical field (at right angles to the ground).

To our surprise, we found a major difference between the two conditions. Chromosomes of the cancer cells in the animals exposed to the horizontal field were markedly abnormal, while those in the animals in the vertical-field group were totally unchanged from their usual pattern. In the horizontal-field group we found chromosome breaks, exchanges of portions of chromosomes from one to another, formations of ring-shaped chromosomes, and tiny fragments that had broken off from other chromosomes.

The production of such severe chromosomal abnormalities usually results in an inability of the cells to divide and ultimately leads either to their death or to the production of a new line of mutant cells. We investigated these possibilities and found that the cancer cells with the marked chromosomal abnormalities died because they were unable to replicate. There were obvious clinical implications from this study, but we were unable to follow up on it, and the experiment was never repeated.

It appears that exposure of a dividing cell to an inhomogeneous DC magnetic field causes a physical force to be exerted on the chromosomes or on one of the other microscopic structures associated with mitosis, resulting in structural abnormalities in the chromosomes.

Complex electronic-resonance effects operate at the level of individual ions or molecules and can occur even in the absence of cells. The effect at the cellular level is dependent upon the organization and the specific type of the cell within which the field effects are produced. We have just begun to explore the effects of fields on cells from these viewpoints.

While the effects of complex electronic-resonance phenomena at the level of ions and molecules may be significant, the resultant change in the function or structure of the cell may be far more important. The effects of field exposure at the level of the total organism will be the sum of the effects at the molecular level, the cellular level, and the level of the specific organs designed to be sensitive to the Earth's normal electromagnetic-field environment.

Field effects on entire functioning organisms are, therefore, a cascade of changes that finally result in many different structural, functional, and behavioral alterations—such as the new diseases that are emerging today. These are the subject of the next chapter.

The data obtained in the past few years indicate very clearly that we must now include the Earth's normal geomagnetic field as an environmental variable of great consequence when we deal with the basic functions of living things. In my opinion, this knowledge is probably the single most important discovery of the century. It provides us with a key to the mechanisms by which all electromagnetic fields produce biological effects, and it may enable us to determine more accurately the risks involved in our technological uses of such fields. More importantly, it opens a door to a greater understanding of life processes, similar to that which was opened by William Gilbert in 1600 when he began the scientific revolution. The subsequent discoveries of the seventeenth and eighteenth centuries gave us our present world. The new discoveries that are linking us to the Earth's magnetic field can give us yet another world, if we explore them properly.

CHAPTER ELEVEN

THE NEW PLAGUES

Where did you come from, baby dear?
Out of the nowhere into here.

Nursery rhyme

*I*n the preceding chapters, the evidence for harmful biological effects from exposure to abnormal, man-made electromagnetic fields was discussed from the point of view of laboratory and epidemiological data. This chapter will present the *medical* evidence relating these fields to recently occurring significant changes in the spectrum of human disease.

We are currently facing diseases that were unknown just a few years ago. We are also seeing disturbing increases and alterations in certain diseases we believed we had conquered. While the AIDS epidemic commands the headlines, there are other illnesses of equal importance—such as Alzheimer's disease, cancer, and developmental defects—that have not received the same amount of attention. The new scientific paradigm we have been considering may provide some clues to these new plagues, some informed speculation, and some indication of the directions we should take.

The global environmental alteration brought about by our use of electromagnetic energy has exposed all living organisms, from viruses to humans, to novel energetic fields that never before existed. We have seen how these fields interact with sensing systems designed to perceive the normal electromagnetic environment, producing abnormal bioeffects that result in changes in the incidence of diseases. This interaction appears also to have been the origin of some new disease states and of a number of unexpected changes in the characteristics of some preexisting disease states.

In theory, a disease that suddenly appears out of nowhere may be caused by a genetic change in a preexisting microorganism (a bacterium or virus) that creates new pathological characteristics. Or, it may be caused by major declines in resistance to disease, making people susceptible to attack by microorganisms previously incapable of producing disease. In practice, things are not quite so simple. Hippocrates was correct in believing that the clinical characteristics of a disease are the result of the combined actions of both the disease-producing agent and the patient's body. When we're facing a new disease, it's often difficult to tell which is which.

ELECTROMAGNETIC-HYPERSENSITIVITY SYNDROME

When I first acquired some public notoriety as a researcher on the biological effects of electromagnetic fields, I began receiving a trickle of letters from persons who insisted that they were highly "allergic" to such fields. Some said they had even moved to remote, rural areas that are free of most electromagnetic fields. I must admit that for the first few years I was highly skeptical of such claims and thought that the conditions must be purely psychological in origin. But during the past five or six years, the trickle of letters has turned into a flood that I can no longer ignore. Fortunately, other medical scientists have become interested in this specific condition and involved in establishing its diagnosis.

Dr. William Rae, a former surgeon from Texas, discovered his own sensitivity to electromagnetic fields while working in the modern operating room. As medicine became a technology, the operating room became a home to more and more electrical devices. Today, in my opinion, it could well be classified as a hazardous environment. By eliminating other sources, Dr. Rae determined that his allergic and neurological symptoms were caused by the electromagnetic fields in the operating room. He subsequently discovered that he was not alone in his hypersensitivity, and that there was a growing population of patients with the same condition. These people are typically told by their physicians that their symptoms are "all in their minds" and that they should seek psychiatric care.

Rae became outraged about this situation, and he established a clinic to deal with this problem as a real entity. His Environmental Health Center in Dallas, Texas, is probably the best-equipped clinic of its kind in the country. The patients are tested through exposure to a spectrum of electromagnetic fields in such a fashion that they are unaware it is being done. In most patients, a consistent sensitivity to certain specific frequencies can be found and quantified through objective measures of the activity of the autonomic nervous system. In this way, Rae has proved that the electromagnetic-hypersensitivity syndrome is a real clinical entity.

People with this syndrome have a number of characteristics and symptoms in common. The following case history is typical.

Mary M. (not her real name) had worked for an international company for many years as a computer supervisor. She enjoyed her work and had had no medical problems of note until she was asked to try out a different make of computer that the company was considering using. The machine seemed perfect—it was easy to use as well as fast and powerful, and she enjoyed working with it the first day. She went home that night with a minor headache, which subsided with an aspirin. Returning to work the next day, she used the new machine for less than an hour, and the headache returned. She took another aspirin and wondered whether she was "coming down with something." As she continued to work with the new computer, she became nauseated and dizzy, and the headache did not go away.

Mary then went to the dispensary, where she was told that she had a slight fever and was no doubt "coming down with the flu." She took two days off, and she felt fine when she returned to work. But within minutes of turning on the computer, she experienced the return of the nausea, dizziness, and headache. A short time later she began to experience severe fatigue, an inability to concentrate, and difficulty with her vision. As she continued to work, the symptoms became worse, until finally she could not continue. She began to think that perhaps something was wrong with the machine, and she mentioned this to the dispensary staff before going home again. When she reached her home, she noted that her face and the exposed portions of her neck and chest were noticeably reddened. This time she took a full week of sick leave; on returning to work she went straight to the dispensary so that the doctor could see that she was quite well before she used the new machine. She was told that in her absence, the manufacturer had checked her machine and found that it was operating normally and not producing a harmful field.

As she opened the door to her area, she felt as though she "had walked into a blast furnace." The whole room had been equipped with the new machines, and her staff was busily working with them. She stayed for only a few minutes, during which time she became extremely ill and had to leave. This time the doctor asked if she were

having any emotional or personal problems of any kind, and he
suggested that if so, she see a professional about them. Mary refused to return to work, and she left for home.

She then noticed that her TV and stereo produced the same
symptoms in her if she was within a few feet of them. Over the
next few weeks, her condition gradually worsened, until even
using the telephone made her ill. She also developed what appeared to her to be "allergies" to sunlight and to the smells of
such things as laundry bleach and perfumes, all of which made
her feel nauseous and dizzy.

The skin rash reappeared, and Mary consulted a dermatologist,
who told her that the electromagnetic radiation from the computer
was the culprit and that he had seen similar cases. He recommended
that she go away for a few weeks to some very rural area and see
whether she recovered. His other patients had sometimes been able
to return to work after such a respite from electromagnetic fields.

Mary took his advice, and she did get better. However, when
she returned to the city, her symptoms returned. She never returned
to her old job, and she now lives in a very rural area of a foreign
country, where she is fine.

One final note on this case is that the computers are no longer
in use by the company, which refuses to discuss the situation.

Almost all of the patients who have consulted me for this condition
experienced a similar abrupt onset of symptoms. Computers were
not always the common factor, but exposure to a novel electromagnetic field of some kind was always the inciting cause. These people
experienced the same symptoms and became sensitive to many common devices that had never before produced any symptoms in them
(for example, TVs, computers, stereos, fluorescent lights, telephones, electric heaters, high-voltage power lines, and electronic
security systems).

Within the past year, automobiles have joined the list of devices
that can produce the initial sensitizing field exposure. These are
always the new, computerized models, and the patients find that
they are quite able to drive the identical make and model if it is not
computerized.

As with any medical condition, there is a range of severity. With repeated exposure some patients develop severe neurological responses, with such symptoms as confusion, depression, decreased memory, sleep disturbances, and even convulsions or grossly abnormal behavior. Others never progress beyond mild symptoms when exposed to the original novel field. In general, there is little that can be done for any of these patients, beyond advising them to move to a location where the fields do not produce symptoms in them.

At present, the mechanisms causing a hypersensitivity to novel electromagnetic fields are unknown. However, it would appear that there is a direct effect upon the nervous system and that the immune system is very likely also involved.

The incidence of this syndrome is increasing. Whether this increase is due to the rising number of new devices emitting novel fields, to the overall increase in field intensity from all sources, or to a change in the sensitivity of the human population is as yet undetermined.

CHRONIC-FATIGUE SYNDROME

Fatigue may be produced by a variety of conditions, ranging from physical exertion, to relatively minor virus infections, to serious infectious diseases and cancer. In 1982, a new syndrome was reported that was at first thought to be either chronic mononucleosis or chronic Epstein-Barr virus infection. It was characterized by severe fatigue, sore throat, tender lymph nodes, mild fever, an inability to concentrate, depression, and mental confusion.

Since then, professional interest has increased, along with reports of a rapidly increasing number of cases. Some investigators believe that this condition is not a new disease but is simply a "fad" produced by publicity. Attempts to track down the causative agent have revealed only that it is not mononucleosis or Epstein-Barr virus. In its most severe form, this condition is prolonged, often lasting for several years, and severely debilitating. In some patients the major symptoms are neurological, with severe disturbances verging on psychotic. Recently, the Centers for Disease Control declared the condition to be a real clinical entity and proposed formal criteria for its diagnosis.

However, no treatment, including the antiviral drug acyclovir, has been found to be effective. In one acyclovir study, done at the National Institutes of Health, matched groups of cases were given either acyclovir or a placebo. The results were identical. However, the investigators (headed by Dr. Steven E. Strauss of the Medical Virology Section) found an interesting association between the results of psychological tests and the changes in a patient's symptoms. "Significant improvement [was seen] in levels of anger, depression, and other mood states correlated with overall clinical improvement." In other words, if the patients *believed* they were getting better, they actually became clinically better.

❋

The hypersensitivity syndrome and the chronic-fatigue syndrome share a number of common characteristics. Both are newly described conditions, both begin rather abruptly with a flulike condition, and many of the symptoms—particularly those associated with the central nervous system—are identical. The major difference is in the reported sensitivity to electromagnetic fields. However, some patients who were diagnosed as having primarily chronic-fatigue syndrome have consulted me because they definitely felt worse when exposed to TVs and other devices, even though they did not consider this to be a major aspect of their illness. Some patients with chronic-fatigue syndrome feel improved when they are in rural areas, away from sources of electromagnetic fields.

Finally, the chronic-fatigue syndrome has been found to be widespread in the electronics industry, particularly in Silicon Valley in northern California. According to Dr. James Cone, chief of Occupational Medicine at San Francisco General Hospital, "These are not people with typical allergies, but people who appear to have some kind of neurological or immune-system problem." More than 100 Silicon Valley employees have filed lawsuits against their companies for the condition.

While by no means proven, the possibility remains that both conditions share the same basic causative factor—exposure to abnormal electromagnetic fields. The differences in the syndromes are that in one type there is overt hypersensitivity to the fields, while in the other this factor is either lacking or is much less evident. Both, however, share the general characteristics of involvement of the central nervous system and the immune system. If the present trend continues and the percentage of the population developing either of these conditions increases, it could become a public-health problem of some magnitude.

AIDS

It now appears quite clear that the causal agent of acquired immunodeficiency syndrome (AIDS) is a new virus called the human immuodeficiency virus, or HIV. However, there is no question that other factors, such as general resistance to disease, behavior, and lifestyle, also play a role in the actual disease syndrome. AIDS came out of nowhere, and it is not at all clear what its antecedents were. In 1980, *Time* magazine reviewed the "new diseases" and the role of the Centers for Disease Control in combatting them. At that time, the mystery ailments were limited to Legionnaire's disease and toxic shock syndrome. Two years later, AIDS appeared. Although its cause was then unknown, its virulence and mortality rate indicated that it was a public-health problem of great importance. This, of course, has been revealed by its subsequent spread and our failure to control it.

The HIV virus shares some attributes with a number of other viruses that cause various leukemias in human beings and animals, but its clinical characteristics are dominated by its attack on the immune system, particularly the T-cell lymphocytes. Dr. Julie Overbaugh and her colleagues at the Harvard School of Public Health have worked with a cat leukemia virus that is capable of producing either leukemia or an acute immunodeficiency syndrome similar to human AIDS. Using molecular-cloning techniques, they were able to produce a mutation of the virus that produced rapidly fatal immunodeficiency syndrome in young cats. They concluded that "a subtle mutational change would convert a minimally pathogenic virus into one that would induce an acute form of immunodeficiency."

It would appear possible that the present HIV virus is the result of such a mutation of a preexisting virus that was originally nonpathogenic, or minimally pathogenic, to people. Because abnormal electromagnetic fields have been shown to be able to cause genetic mutations, perhaps this mechanism was involved in the mutation of a nonpathogenic virus into the HIV virus. While this may seem at first to be farfetched, it is not beyond the bounds of possibility.

It would be quite unusual if clinical AIDS disease was simply the result of infection with the HIV virus, regardless of the state of the human host. It is likely that other factors, such as the status of the immune system, increase the chance that an infected person will develop the overt clinical syndrome. Therefore, the depressant effect on the immune system of chronic exposure to abnormal fields may also be involved. In a recent paper, Dr. Daniel B. Lyle of the Jerry L. Pettis Memorial Veterans Hospital in Loma Linda, California, reported that exposure of human T-cell lymphocytes in culture to a low-strength 60-Hz electric field for forty-eight hours significantly reduced their cytotoxic ability against foreign cells. Lyle's report is the first to link such fields directly with the cells of the immune system. This finding raises the question of a relationship between the depressant effect of power-frequency fields on T-cell function and the susceptibility of these cells to infection with the HIV virus.

The orthodox physicians in charge of the response to the AIDS epidemic are locked into the chemical-mechanistic model, which involves the idea of a "magic bullet" that would kill the AIDS virus. The truth is that we have no chemical agent that is effective against *any* viral disease, let alone one so sophisticated and genetically changeable as HIV. The possibility that an effective antiviral chemical or vaccine will soon be developed appears remote.

It is time to stop thinking about AIDS as being simply the result of infection with the HIV virus and to begin to consider the possibility that these electromagnetic factors, or similar ones, might play a predisposing or causal role in the total syndrome. We also need to begin an organized study of those techniques of energy medicine that are known to enhance the operations of the immune system. Considering the fact that there is no known curative or life-prolonging treatment available for AIDS, I see no ethical consideration that would preclude the use and evaluation of these techniques.

AUTISM

The autistic-child syndrome was first described in 1943 as an apparently "inborn disturbance of social development," in which children from very early infancy onward showed "profound social disinterest" and "bizarre responses to the environment, unusual behavior, and marked disturbances in communication." Thereafter, professional interest in the condition increased, as did the number of cases.

Several theories were proposed for the cause of autism, including overt psychosis (schizophrenia), behavioral disturbances, and others. Later, autism was clearly differentiated from schizophrenia and was often found to be the result of a "variety of insults to the developing central nervous system—e.g., congenital rubella" (maternal measles infecting the brain of the fetus) or to be associated with mental retardation. At present, the disease is considered to be the result of an actual biological injury or insult to the brain occurring during fetal development. A number of studies have looked for some specific pathological lesion in the brain, and while several such lesions have been reported, their occurrence appeared to be almost random among the cases studied.

In 1988, Dr. E. Courchesne of the Neuropsychology Research Laboratory at Children's Hospital Research Center, San Diego, reported finding a specific pathological lesion of the brain in fourteen out of eighteen autistic patients. The brains of these children were examined using high-resolution magnetic-resonance scans, and a specific anatomical portion of the cerebellum was found to be significantly smaller and less developed than the same structure in normal children. (The cerebellum is a small, distinct portion of the brain located in the lower back of the head. Its major function is generally considered to be the coordination of motor movements; however, this is probably a gross oversimplification.) This specific lesion in the cerebellum appears to be directly related to the clinical autistic syndrome.

In the early 1980s, Dr. Hans-Arne Hannson of the Institute of Neurobiology at the University of Göteborg, Sweden, began investigating the effect of exposure to both microwave and power-frequency fields on the brains of newborn experimental animals. In the case of microwaves, he reported that brief exposure to thermal levels resulted in damage to nerve-cell structures that became visible, as a latent effect, only two to four months following the exposure. The nerve-cell damage was visible in the brain, retina, optic nerve, and cerebellum. In animals exposed to 50-Hz power-frequency fields, Hannson found that pathological anatomical changes were produced specifically in the cerebellums of newborns.

More recently, Dr. Ernest N. Albert of the George Washington University Medical Center found essentially the same pathological changes in the cerebellums of newborn rats that had been exposed to low-power microwave at one or six days of age.

At this time, we cannot be sure whether the actual lesions in the cerebellums of autistic children match those in the cerebellums of animals that were produced by exposure to either power-frequency or microwave fields. However, the apparent onset of autism as a clinical condition in the early 1940s does coincide with the marked increase in our usage of electromagnetic energy. *The fact that both autistic children and the experimental animals have been found to have lesions in the same specific portion of the brain is a remarkable coincidence, which must be studied in depth.*

Autism is a devastating disease, and only 1 to 2 percent of its victims are ever able to achieve a totally independent personal and occupational existence. It is vitally important to determine whether this disease is the result of exposure to abnormal electromagnetic fields, whether during the final stages of fetal life or the early newborn stage.

FRAGILE X SYNDROME

One of the chromosomes identified in the human cell is the *X* chromosome, so named because of its shape. The *X* is one of the sex chromosomes, along with the *Y* chromosome. The male-female gender decision in each individual is determined by the ratio of *X* and *Y* chromosomes in the fertilized egg.

Twenty years ago, a syndrome characterized by various degrees of mental retardation, behavioral disturbances, and certain anatomical characteristics was found to be related to an abnormality in the *X* chromosome. The *X* chromosomes of these patients show an apparent separation of a fragment of one of the limbs of the *X*—hence the term Fragile *X* syndrome. It appears that the defect occurs during early mitoses of the fertilized egg and is caused by a failure of the chromatin (the aggregated DNA that becomes the separate chromosomes) to "glue" together properly at this site.

It is now recognized that Fragile *X* is second only to Down's syndrome as most common genetic defect in this country. The incidence of Down's syndrome is about 1 case per 1000 births, while Fragile *X* occurs in between 1 per 1000 and 1 per 1500 births. Strangely, not all persons with fragile *X* chromosomes have the syndrome, and there is some evidence that the condition is passed from mother to son. Research is progressing in unraveling the complexities of this important condition, and the most recent finding is of great interest.

Immediately following the report of Dr. Courchesne on the defect in the cerebellums of autistic children, Dr. Alan Reiss of the

Kennedy Institute at Johns Hopkins University examined the brains of patients with the Fragile X syndrome. He found a very similar defect in the cerebellums of these patients.

Perhaps this cerebellar defect is but one part of a larger neurobiological abnormality. The anatomical abnormalities found by Courchesne and Reiss, and the production of the same abnormalities in animals through exposure to abnormal electromagnetic fields that was observed by Hannson and Albert, raise some disturbing questions about the future.

Many newborns in this country are exposed to abnormal electromagnetic fields in hospital nurseries and intensive-care units for premature infants. These fields are generated by the extensive use of high-technology monitoring and heating devices. In a specific instance, jaundice has become a frequent problem in newborns. It is commonly treated with phototherapy, which is exposure to a source of intense white light—usually, fluorescent bulbs which produce a high-strength electromagnetic field (see chapter 12). This treatment is effective and necessary, because the chemical causing the jaundice (bilirubin) is toxic to brain cells. The intense light causes a change in the chemical structure of the bilirubin, enabling it to be excreted in the urine.

The only side effect of the therapy that has been considered is possible eye damage, so babies are commonly equipped with eye shields during the treatment. However, it has been noted that infants treated in this fashion have lower-than-normal levels of blood calcium. In 1981, doctors David Hakanson and William Bergstrom of the State University of New York Medical Center, Syracuse, studied the effects of phototherapy on newborn laboratory animals. They determined that the lower calcium was due to an effect upon the pineal gland, resulting in inhibition of melatonin secretion. The phototherapy evidently has a direct effect upon this important structure within the brain, raising the possibility that other effects of greater importance may occur in both animals and humans.

The cells of the brain and central nervous system appear to be particularly sensitive to abnormal fields. This sensitivity may be expressed as disturbances in the function of adult cells, such as Adey's demonstration of the Ca^{++} efflux or the pathological changes leading to brain tumors (see chapter 8). The developing central nervous system of the fetus or the newborn is particularly sensitive. For the first few months of life, the brain of the newborn infant is said to be "plastic," because it is rapidly changing and making new connections and anatomical arrangements. Exposure of the brain to abnormal electromagnetic fields at this time may result in either the production of abnormal connections or actual permanent anatomical changes, with tragic results.

As we saw earlier, the pineal gland is the principal structure in the brain that is directly sensitive to the Earth's magnetic field. As a result, it functions abnormally when exposed to abnormal fields. Because the pineal produces a host of psychoactive chemicals (such as melatonin, dopamine, serotonin, and others), its abnormal functioning during this time of brain plasticity may result in a variety of transient or permanent neurological and behavioral abnormalities. Some of these effects may be subtle and not yet identified; others may be related to such disturbances as slower learning ability, demonstrated by Wolpaw, and the phenomenon of sudden, unexpected infant death.

SUDDEN INFANT DEATH SYNDROME (SIDS)

The tragic phenomenon of the sudden, inexplicable death of apparently healthy sleeping infants has been recognized for a long time. I have been unable to find any data on changes in the incidence of this condition; however, many of my pediatrician colleagues share my opinion that it is increasing. Many studies have been undertaken to determine the cause, but to date no cause has been found.

Recently, Dr. William Sturner, chief medical examiner for the state of Rhode Island, presented some interesting findings on forty-five infant deaths, eighteen from SIDS and twenty-seven from other established causes. Using an ultrasensitive method, he measured the infants' levels of melatonin, a neurohormone produced by the pineal gland. He found that the levels of melatonin in those infants who had died of SIDS were significantly lower than those in the other infants.

The brain levels of the hormone in the SIDS cases averaged 15 picograms per milliliter (pg/ml), compared with 51 pg/ml in the control group. Blood levels of the hormone in the SIDS cases averaged 11 pg/ml and 35 pg/ml in the control group. The infants who had died of SIDS had a much lower-than-normal overall level of the hormone, which may have resulted in depression of respiratory controls to the point at which these infants stopped breathing. At this time, it is not possible to say whether the pineal in SIDS infants is overactive or underactive, but the relationship between its function and environmental electromagnetic fields would seem to indicate that some abnormal, man-made field might be responsible.

Dr. Cornelia O'Leary, a fellow of the Royal College of Surgeons in England, has been studying the possible relationship between SIDS and abnormal electromagnetic fields. Recently, she reported that eight such deaths occurred in one weekend (four of these within a single two-hour period) within a radius of seven miles from a top-secret military base where a powerful new radar unit was being tested. This certainly indicates that a link is possible.

CHANGES IN PREEXISTING DISEASES

Alzheimer's Disease

If fetal or newborn brain cells are particularly sensitive to electromagnetic field exposure, perhaps aging brain cells are also especially susceptible. If this is so, exposure to these fields may be associated with specific disease patterns among older people. Alzheimer's disease is a true disease, not simply dementia resulting from arteriosclerotic disease of the brain in the elderly. Specific pathological changes are found in the brains of people afflicted with this condition. (Patients with Down's syndrome often develop the clinical signs of Alzheimer's disease at about age forty, if they survive to that age; in addition, the pathological lesions associated with Alzheimer's are found in the brains of all Down's patients of this age, whether they display the clinical signs or not.)

There is some genetic linkage in Alzheimer's disease, with some families showing a predisposition toward developing it. However, as Dr. Robert Katzman of the Department of Neuroscience at the University of California, San Diego, has stated, "Even in identical twins, environmental or metabolic factors must have a role in determining the onset of this disorder." There have been a number of reports indicating that Alzheimer's patients have a chromosomal abnormality similar to that occurring in people with Down's; however, this correlation was not found by Dr. Peter St. George-Hyslop of the Neurogenetics Laboratory at Massachusetts General Hospital. The situation remains unclear. If further studies indicate that a subtle genetic change is actually responsible for Alzheimer's disease, one may well ask whether this change is environmentally produced, possibly by exposure to electromagnetic radiation.

Dr. Sam Koslov, director of the Applied Physics Laboratory at Johns Hopkins University, recently presented preliminary results of a microwave-exposure study on chimpanzees. At an EPA-sponsored public meeting on environmental electromagnetic fields, Koslov reported a possible link between Alzheimer's disease and microwave exposure. Koslov, who has been associated with the electromagnetic bioeffects field since the 1940s, was studying the effects of microwave exposure on the eyes of chimpanzees. The animals were exposed repeatedly to low-level, nonthermal microwaves, and their eyes were continuously examined. As the study progressed, one animal began to demonstrate the classical clinical signs of Alzheimer's disease. On autopsy, this animal's brain showed the typical pathological picture associated with Alzheimer's disease. Koslov stated that he was revealing this information to the public out of frustration with the lack of funding for the whole area of electromagnetic bioeffects.

The incidence of Alzheimer's disease is increasing. At present, it is estimated that there are 2 million cases in the United States. Because the increase is generally attributed to an "aging" population, there have been no studies done on its possible environmental relationships.

Parkinson's Disease

Parkinson's disease is prominent among those "normal" diseases that have increased in incidence over the past few decades. Not only is the total incidence rising, but in addition the numbers of people who get the disease before the age of fifty may have increased by as much as 50 percent over the past thirty years, according to Dr. Donald Calne of the University of British Columbia. In an interview with Roger Lewin of the journal *Science,* Dr. Calne also stated, "It is difficult to avoid the conclusion that there is an environmental risk factor in the disease process which is becoming more common." The Canadian researchers have also found that, like Alzheimer's, Parkinson's disease seems to have a familial tendency. While no direct linkage with electromagnetic fields has been investigated, the effect of this environmental factor on the central nervous system indicates that such a search might prove fruitful.

Cancer

I have already indicated that the overall rate of cancer in the United States is steadily increasing, despite improvements in diagnosis and treatment and declines in the rates of some of the more common types of malignancies. Part of the explanation is that this overall rise in incidence is derived from much greater increases in the incidence of cancers of tissues in which there is a high rate of cellular multiplication. In that group is a particularly interesting cancer, melanoma of the skin.

In a recent review article, Dr. Mark H. Green and his colleagues at the National Cancer Institute wrote that "the incidence of cutaneous, malignant melanoma is rising rapidly throughout the world. The most recent data from the National Cancer Institute's Surveillance Epidemiology system reveal an 80 percent increase in the incidence of melanoma in the United States between 1973 and 1980." Dr. Thomas B. Fitzpatrick of the Harvard Medical School has likened the increase to an "epidemic," and he has stated further that this type of cancer is not only the fastest growing of all cancers but is also affecting a younger age group each year.

Because it is known that overexposure to sunlight is one of the causes of melanoma, there has been much recent speculation as to a possible link between its increased incidence and the decline in atmospheric ozone. This seems unlikely, however, because the increased rate of melanoma predates the decline in ozone and because the "ozone holes" are limited to the arctic regions, with no measurable decline found in the ozone layer over the United States.

A clue as to a possible causative factor for this disturbing trend may be found in the one place in this country that has the highest incidence of melanoma, the Lawrence Livermore National Laboratory (LLNL) at Livermore, California. The laboratory is deeply involved in the design and testing of new and exotic weapons. In 1977, the cancer-epidemiology branch of the California Department of Health Services began a study of what appeared to be an excess of malignant melanoma of the skin among LLNL employees. The results of the study revealed a three- to fourfold excess of such cases over what would be expected. The excess was limited to LLNL personnel and was not duplicated in the surrounding civilian communities (which were then experiencing the slower nationwide rate of increase in the incidence of this type of tumor). In other words, the personnel of LLNL were experiencing a much higher incidence of melanoma on top of the general increase in the rest of the country.

In 1985, doctors Peggy Reynolds and Donald Austin, the directors of the study, reported the results. They noted that while melanoma at LLNL was elevated above normal, so were the rates for tumors of the salivary gland, colon, and brain. However, the rates of incidence for these other tumors were not too dissimilar from those in the surrounding communities. Only malignant melanoma stood out as special to LLNL. Reynolds and Austin emphasized that the LLNL personnel showed no excesses of the types of tumors that would have been expected if ionizing radiation (X-ray, nuclear radiation, and so on) had been the causative factor. Therefore, something else in the LLNL environment had to be the cause for the gross excess of melanoma, but they had been unable to find that "something else."

While no data were available on exposure of LLNL personnel to nonionizing electromagnetic fields, the nature of the work being done indicates that such exposure might have been high. For example, when I was consulted by LLNL on the possible hazards of personnel exposure to DC magnetic fields, I learned that some individuals were routinely exposed to DC magnetic fields as high as 1400 gauss for the entire workday, and that the safety standards were set at 2000 gauss for the body's trunk and 20,000 gauss for the extremities. These are truly high-strength fields, and we have no prior experience with the effects of prolonged exposures to them. In view of this inadequate data base, the origin of the "safety standards" is highly questionable. The situation regarding DC magnetic fields may represent the exposure situation to electromagnetic fields in general at LLNL.

In view of the stimulatory effects of abnormal electromagnetic fields on cancer cells and their ability to alter the genetic apparatus of dividing cells, this peak of the epidemic of melanoma at the Livermore laboratory becomes most interesting. It would seem prudent to assess the actual field exposure of the LLNL personnel and to conduct a prospective epidemiological survey. Unfortunately, the classified nature of the work done at LLNL precludes such an investigation.

Mental Diseases

In 1980, the National Institutes of Mental Health
(NIMH) began a long-range study of the incidence of mental disease
in this country. In October of 1984, Dr. Darrel Regier of NIMH
announced at a press conference that early results indicated that
about 20 percent of the total U.S. population had mental disorders
severe enough to require psychiatric treatment. More significantly,
people under the age of forty-five had a rate of mental disorders that
was twice as high as the rate for those over forty-five. This alarming
trend indicates that the incidence of mental disease is rising
markedly. The increase is not in true psychoses (such as schizophrenia and manic-depressive states), but in neuroses, such as depression, phobias, antisocial personality traits, alcoholism, drug addiction, and suicide.

The increase in adolescent suicide rates is particularly disturbing. In a 1986 editorial in the *New England Journal of Medicine*,
Dr. Leon Eisenberg of the Harvard Medical School wrote that "between 1950 and 1977, the suicide rates for 15-to-19-year-olds rose
precipitously—fourfold in males, and almost twofold in females."
Further, he wrote, "A suicide rate higher during adolescence for any
cohort [group of persons born during any five-year span in time]
than for the previous cohort predicts that the rate will be persis-
tently higher for that cohort as it ages."

The message of the first quote is self-evident; the second means
that for any group displaying a higher suicide rate, the rate will
remain higher for the rest of the life of that group. Dr. Eisenberg
concludes, "Clearly, adolescent suicide is a major health problem for
the long term, as well as for the short term, and the search for
factors that influence its frequency takes on added urgency."

Over the past three or four decades, some environmental factor
has been introduced that has seriously influenced the basic levels of
mental functioning, perhaps in the entire population. The fact that
the effect is seen most overtly in younger age groups seems to
indicate that the factor is operating during the early stages of life.
There does seem to be a direct link between suicide, depression, and
functions of the pineal gland.

At a 1986 conference on suicidal behavior that was sponsored
by the New York Academy of Sciences, Dr. Marie Asberg of Sweden's Karolinska Institute presented data indicating that a group of

depressed patients with a deficiency in serotonin had a significantly higher suicide rate than a similar group of depressed patients whose serotonin levels were normal.

It seems that there may be two types of clinical depression: one that is produced by simple psychosocial factors, and one that is produced by some external factor that influences the production of these psychoactive chemicals by the pineal gland. In view of the known relationship between the pineal gland and magnetic fields, it is advisable that the search for the responsible factor include an evaluation of the effect of abnormal electromagnetic fields, particularly in the early years of life.

WHERE DO WE GO FROM HERE?

While some of the preceding discussion is speculative, it is informed speculation, based upon a wealth of currently available data that indicate a plausible link between abnormal, man-made electromagnetic fields and these changes in the spectrum of diseases. Taken together with the data contained in chapter 8, these new diseases and the disturbing changes in preexisting diseases constitute a major health problem for the nation. The medical profession needs to begin to deal with them in a more effective manner.

There is a need for a change in the basic philosophy of medicine in its handling of these challenges. Because we are dealing with the effects of electromagnetic energy on the energetic systems of the body, the new scientific paradigm that lends support to the basic concepts of energy medicine provides a good place to start. Two pathways should be followed. First, epidemiological studies must be expanded to include the possible causative effects of environmental electromagnetic fields. Second, the therapeutic possibilities of utilizing the body's own energetic systems to treat these new conditions—either alone or supported by appropriately designed and applied external fields—must be simultaneously explored.

The growing tendency to integrate the concepts of energy medicine with those of orthodox medicine must be encouraged. If those who are totally committed to the orthodox, chemical-mechanistic paradigm are solely in charge, much valuable time will be lost. To illustrate this point, the National Cancer Institute recently announced that it will begin a full-scale study of the possible relationship between exposure to power-frequency fields and childhood cancer rates. While this sounds appropriately responsible, and even hopeful, the problem is that many of the top-level scientists at NCI have over the past two years testified in legal proceedings that power-frequency electromagnetic fields have no biological effects and that exposure to them is absolutely harmless. It would appear highly unlikely that an unbiased study could be conducted under these circumstances.

While the AIDS epidemic is often referred to as a medical crisis from the point of view of the growing number of victims and the mounting costs of care, I believe that this is far too narrow a perspective. At a time when cancers of many types are increasing in incidence and virulence; the number of children born with developmental defects is increasing; 50 percent of the population below the age of forty-five have serious mental problems; major neurological degenerative diseases are found in both the very young and the very old with increasing frequency; and we have no effective treatments for any of these conditions, AIDS may be seen as simply the tip of the iceberg of a much greater problem.

What will be the effect upon our institutions and culture if these trends continue to increase in extent? Clearly, we are facing a crisis of major proportions, one that is all the more critical because it has not been recognized as such by the agencies responsible for dealing with it.

The public must play a major role in resolving this problem. Without a public that is educated as to the real extent of the crisis and that demands effective action of its officials, nothing will be done until the situation becomes so critical that remedial actions may be impossible. In addition to the environmental situation, people also have responsibilities for their own personal health and safety. None of these responsibilities can be delegated to others.

CHAPTER TWELVE

THE RISK AND THE BENEFIT: WHAT YOU CAN DO

Before the dams were built, the river could speak. People would talk to the river. The trees and the animals could talk. The spirits of the hunters who died on the land walked the riverbanks with the spirits of the animals they had killed. These things must have been true, because the people believed them.

[After the land was flooded,] it was like a killing. They have drowned things that were living. What are they doing to us? What are they doing to our garden?

From an interview with Joe Bearskin of the Cree Indians by Bruce DeSilva of *The Hartford Courant*

*T*he James Bay Hydroelectric Power Project is the largest construction project ever undertaken by human beings. In the wilds of northern Canada, rivers have been made to flow backwards, and thousands of square miles have been flooded. Eventually, the ancestral home of the Cree Indians will be destroyed—all to provide electrical power for the East Coast of the United States. The electrical and electronics industries are constantly expanding their installations and applications. Their representatives tell us that if they do not do so, our future standard of living will suffer.

The possible health effects of electromagnetic fields have become sufficiently well known that I am frequently asked what private citizens can do to protect themselves. Generally, this question arises in regard to the proposed construction of a new electric-power transmission line, microwave-relay station, or other such installation. Unfortunately, the answer is not simple; there are actually several actions that can be taken, all of which are interrelated.

First, we need to determine the relative risks associated with our use of electromagnetic energy, personally and collectively, versus the benefits we derive from that use. Second, there are personal actions that individuals may take relating to the devices they use in their homes or offices. Third, there are actions that we may take collectively, in concert with our neighbors, in regard to a common problem concerning environmental electromagnetic fields.

RISK/BENEFIT RATIO

We do not live in a risk-free society. Many of the beneficial technological devices we use are potentially dangerous, yet we continue to use them because we believe that the benefit exceeds the risk. The best example is the automobile, which is involved in a great many injuries and deaths each year. However, the structure of our society is such that the majority of citizens feel that a car is essential. Most of us take the usual precautions against accident and injury, and we accept the risk that they may occur. In this case, each individual makes his or her own risk/benefit analysis.

I have found that there are many misconceptions about the health risks associated with electromagnetic fields. The most common is the assumption that such fields arise primarily from external environmental sources. Most people are unaware of the substantial risks associated with the use of many devices in their own homes. It is not logical to fiercely oppose the installation of an electric-power transmission line while using an electric blanket every night.

Because our global society runs on electromagnetic energy, there really is no place to hide. Even in the most remote mountain valleys one is exposed to some level of the ubiquitous power frequencies of 50 or 60 Hz, as well as to shortwave radio waves reflected back from the ionosphere. Resolution of the problem of global electromagnetic pollution of the environment will require a concerted international effort. However, as individuals we have some control over the electromagnetic devices that we use every day. All we need is an understanding of these devices and of the comparable risks associated with each, and a knowledge of how to evaluate the risks in a rational manner.

Dose-Rate Considerations: Electric Razors and Electric Blankets

The one basic concept that we need to apply is dose rate. For example, the electric razor produces an extremely high-strength magnetic field if it is a line-operated device—that is, if it is plugged into a wall socket rather than battery operated. I have measured 60-Hz fields as high as 200 to 400 milligauss one-half inch away from the cutting edge of a number of different makes of razors. When such a razor is in use, the tissues within a few inches of its surface are exposed to a high magnetic field. Because 60-Hz fields of only 3 milligauss have been shown to be significantly related to increases in cancer rates, the safety of using this device—with a field 100 times greater—becomes questionable.

However, the dose-rate concept must enter the picture. The electric razor is used for just a few minutes once a day, so the total exposure, or dose, to the user is minimal. Electric-blanket field strengths are somewhat lower (50–100 milligauss) but are still within the danger zone. The electric blanket is applied almost as close to the body surface as the electric razor. Because it is used for several hours at a time, the total administered dose is much higher.

Dr. Nancy Wertheimer of the University of Colorado, who published the first epidemiological study on power-frequency fields, has done similar studies on users of electric blankets. She has found that the incidence of miscarriage is much higher among pregnant women who use electric blankets than among those who do not. Of course, this does not mean that other effects, such as developmental defects in the offspring or cancer, are not possible. Miscarriage happens to be an easy and suitable index to use for a preliminary epidemiological study. Obviously, the risks involved in using an electric blanket are quite significant, while using an electric razor is probably safe (I have to say "probably," because there are, as yet, no data on such short-term exposure to high-strength fields).

A comparison of the electric razor and the electric blanket provides a useful example of how one can rationally apply the risk/benefit concept. For both devices there are obvious alternatives that satisfactorily serve the same function. If one takes the risk of miscarriage seriously, the answer is obvious: either use ordinary blankets, or heat the bed with the electric blanket before retiring and then *unplug* the blanket. Do not just turn the switch to off—many electric blankets still produce an electric field if left plugged into the wall socket. The time and effort involved in taking this precaution are minimal, and the advantages are obvious. In this instance, the risk/benefit ratio is clearly tilted toward the risk side.

In the case of the electric razor, no risks have been identified with its use, even though the field strength is high. However, it, too, is easily replaced with the safety razor at the cost of a little more time each day. To be perfectly safe and to eliminate even the theoretical risk, one can substitute the safety razor for the electric razor. (And because I believe the use of an electric razor may increase the risk of stimulating a melanoma among men with pigmented facial moles, I advise against it in this case.)

Obviously, risk/benefit analysis becomes much more complicated when the device associated with the risk is difficult or impossible to replace; the automobile is a good example.

THE AMBIENT FIELD

The ambient field is produced by the local electric-power transmission and distribution network, and it is the level of field strength to which we are constantly exposed. This field is present both outside and inside our homes. The popular idea that aluminum siding will shield the interior of the house from electromagnetic fields is incorrect. Unless every piece is carefully bonded together to make a single unit, which is then well grounded, aluminum siding is useless. Shielding a home from the ambient field is a practical impossibility.

In urban environments, the ambient-field levels often exceed 3 milligauss; in the average suburban home, they range from 1 to 3 milligauss; and in rural areas they are generally less than 1 milligauss. These numbers may be higher depending on the proximity of the home to electric-power transmission lines, power-line transformers, and so forth.

The studies of Wertheimer, Savitz, and others (see chapter 8) indicate that residential exposure to ambient fields greater in strength than 3 milligauss are significantly related to increases in the incidence of childhood cancer. There is good evidence that such fields may also be associated with adult cancers. In risk protection, a factor of ten is generally applied. In this case, that would drop the theoretically safe level to 0.3 milligauss.

Because of practical considerations, I advocate a maximum field strength of 1 milligauss for continuous exposure to 60-Hz fields. This opinion is based on the best evidence currently available. If you are concerned about the strength of the background field in your own home, the only way to determine it is to measure it yourself or have someone else do it for you. When such a measurement is made, all interior field-producing devices should be turned off, and the measurement device should be held five feet away. (Manufacturers of appropriate devices are listed in the Resources section for this chapter.) Because the ambient field is produced outside the home, suggestions on what to do if the field strength in your home exceeds this level will be found later in this chapter in the section on environmental fields.

PERSONAL ACTIONS

In our everyday activities we use a number of devices that give off electromagnetic fields. In doing so, we add that field strength to the overall ambient field strength. Remember, however, that the duration of exposure is important.

Let's look at some common items in the home from this point of view. Some of what follows may surprise you.

Television Sets

The television set is one of the most common items in the American home, and many people seem to spend a fair number of hours per day in front of it. Most people know that a small amount of ionizing radiation is given off by the screen (as X-rays), but few are aware of the amount and distribution of the nonionizing electromagnetic radiation from the entire unit.

The TV set is what we call a broadband radiating source, meaning that it gives off a variety of frequencies, from the 60-Hz power frequency to radio frequencies in the MHz range. Since the picture is made up of separate horizontal lines, each constantly generated from left to right (the raster sweep), every TV set contains a unique circuit (the "fly-back" circuit) that returns the line sweep to the left side of the screen at the end of each line. The fly-back circuit operates in the VLF range, generally around 17 kHz, and this frequency is a major part of the total frequencies given off. When you sit in front of the set, you are being irradiated with a wide range of frequencies, with the strongest components probably being the 60-Hz power frequency and the 17-kHz fly-back frequency.

It is a very common misconception that the electromagnetic field is given off only by the screen itself. Actually, the electromagnetic fields are generated within the circuitry of the entire set. Because the box containing the circuitry is transparent to this radiation, the radiation is given off in all directions from the TV set. The field patterns are not uniform and will differ among different sets, even those with the same screen size.

In general, the larger the size of the screen, the stronger the fields, and the further out from the set they will extend. (The reason for this is that the size of the screen determines the power at which the circuits must work.) Therefore, for large-screen televisions one must sit further away to be exposed to no more than a 1-milligauss field.

One thing to keep in mind is that electromagnetic radiation

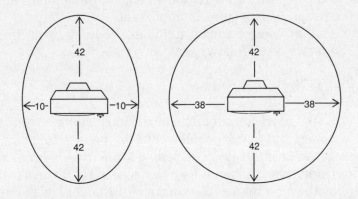

FIGURE 12-1. *The pattern of the 1-milligauss field from two 13-inch TV sets made by different manufacturers. The numbers are in inches, and the figures are not to scale. The set on the left is not necessarily safer, since the video image cannot be seen from the sides, where the field strength is less. In both cases, the minimum distance one must sit from either set in order to be exposed to no more than a 1-milligauss field is 42 inches. Larger-screen sets have proportionally larger fields with 1-milligauss field strengths.*

goes through wood and other usual building materials. If the TV set is placed with its back against an inside wall, radiation will be present in the adjoining room, just as if there were no wall present. Consequently, an infant's or child's bed should *not* be placed against a wall opposite a television set, regardless of the set's field strength.

In 1987, Dr. H. Mikolajczyk and his colleagues from the Institute of Occupational Medicine in Lodz, Poland, presented the results of a study on TV radiation at a conference called "Work with Display Units" in Stockholm, Sweden. The researchers had exposed rats to commercial TV sets, which were placed 30 centimeters (about 12 inches) above the animals. The sets were turned on for four hours each day. Female rats were exposed for sixty days prior to mating and then for sixteen days during pregnancy. It was found that fetal weights were significantly reduced.

Male rats were exposed for thirty-five to fifty days and were then examined. Their testicular weights were found to be significantly reduced. In all animals exposed, concentrations of sodium in the brain cortex, hypothalamus, and midbrain were reduced to below-normal levels. Among growing rats of both sexes, growth rates were significantly retarded. In general, exposure to the electromagnetic radiation from television sets slowed growth, reduced the size of the male rats' testicles, and affected the functions of the brain.

Personal Computers and Video-Display Terminals (VDTs)

Over the past ten years, the personal computer has become almost as popular as the TV set in the American home. In offices, it is now commonplace to see literally hundreds of such machines lined up in rows. Models manufactured prior to about 1982 produced strong broadband radiation. When these models began to invade offices adjacent to commercial airports, their interference with control-tower operations became apparent. As a result, the FCC enacted rules limiting the amount of radiation computers could produce, and the models manufactured since then have less radiation leakage. However, even today, none of the newer models on the open commercial market is adequately shielded.

The radiation pattern is similar to that of the TV set. The major difference in the hazards associated with use is that compared to the

TV viewer, the computer operator must ordinarily sit much closer to the device.*

Almost from the very beginning of the computer revolution, questions were raised about the safety of these devices. These typically were anecdotal accounts of "clusters" of miscarriages among female computer operators. (The term "cluster" refers to a greater-than-average incidence of miscarriage among a group of women working at a single location. By "anecdotal," we mean simply that the events occurred spontaneously in the course of business and eventually were reported by the women in question.) There were no

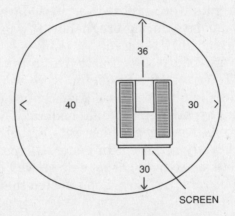

FIGURE 12-2. *The field pattern at 1 milligauss around a Macintosh Plus computer that meets present leakage standards. The figure is not to scale, and the distances are indicated in inches. The asymmetrical field is due to the fact that the power supply is located on the left side of the computer as you face it, and the 1-milligauss boundary extends out further in that direction.*

---

*There is a subtle, but very important, difference between television sets and computers regarding radiation leakage. The radiation given off by a computer contains the information that the computer is processing. It is quite possible to "read" this information from as far away as half a mile, using devices that have no direct connection to the computer itself. This can obviously pose problems to such organizations as the CIA and other such governmental agencies. The problem was code-named TEMPEST, and the term "TEMPEST protected" became associated with computers that were totally shielded from radiating information. The protection involves classified technology and adds approximately $3,000 to the cost of each desktop model. The government spends this kind of money to protect the information, not the operator.

control groups, and no scientific studies were done. Each event was simply handled by a manufacturer's representative appearing in the office, making a few measurements, and assuring everyone concerned that the radiation was far below any established safety standard. Any blame for the miscarriages was placed on "work stress," "poor seating," or "bad lighting."

But the numbers of such clusters multiplied, and in 1985 Robert DeMatteo, coordinator of Occupational Health and Safety for the Ontario (Canada) Public Service Employees Union, collected the data then available and published a book entitled *Terminal Shock.* He reported the identification of eleven such clusters in the United States and Canada, and he raised the question of whether nonionizing radiation was the causative agent. In addition to miscarriages, DeMatteo reported a number of other medical conditions associated with computer use. These included birth defects in offspring of operators, cataracts and other vision problems, disturbances in menstrual cycles, skin rashes, and the general constellation of symptoms associated with chronic stress, such as headache, nausea, sleeplessness, and fatigue.

Many computer models have attached keyboards, so the operator's head is only about fifteen to eighteen inches from the screen. Even with models that have detached keyboards, the screen display is most easily read from this distance. Because the strength of the field drops off with the square of the distance, the closer one's head to the machine, the stronger the field.

In the mid-1980s, scientific studies were begun in Sweden and other European countries using animals exposed to computer radiation under controlled conditions. In 1986, doctors Bernhard Tribukait and Eva Cekan of the Karolinska Institute and Dr. Lars-Erik Paulsson of the Swedish National Institute of Radiation Protection reported that mice exposed to this type of radiation had *five times* the incidence of developmental malformations in their offspring than did control mice. These results were confirmed by a study done the following year by Professor Gunnar Walinder at the Swedish University of Agricultural Science in Uppsala.

Despite these scientific findings, each cluster report concerning humans has been discounted by those in authority. It was not until very recently that any large-scale epidemiological study on computer users was completed. In 1988, doctors Marilyn Goldhaber, Michael Polen, and Robert Hiat of the Kaiser Permanente Health

Group in Oakland, California, reported the results of a study done on a group of 1,583 pregnant women. The results showed that female workers using computers more than twenty hours per week had twice the miscarriage rate of female workers who did similar work but did not use computers. While the number of birth defects in the entire group was too small for accurate statistical analysis, the Kaiser team reported that women using terminals more than twenty hours a week have a 40 percent increased incidence of miscarriage compared with nonusers.

In June 1988, despite the laboratory reports and the Kaiser Permanente study, the International Radiation Protection Association still insisted that "there are no health hazards associated with radiation or fields [from computers or VDTs]." I believe that the evidence belies this optimistic point of view. Obviously, the data indicate that some reproductive risk is associated with the use of personal computers, both at home and in the office. Other risks, such as cancer, have as yet not even been studied.

Some simple steps can be taken to minimize the risks of using these machines. Using a detachable keyboard and moving the computer proper, the display screen, and any peripheral devices (such as disk drives) at least thirty inches back from the keyboard will reduce the field level at the operator's head to an average of 1 milligauss. Any difficulty in reading the screen at this distance can usually be taken care of by wearing corrective eyeglasses.

The advantages of computers have become obvious. The manuscript of this book is being typed on a Macintosh with a detachable keyboard, which is placed thirty inches away from the other components. However, the thirty-inch distance is not guaranteed to be safe, and each person must determine his or her own risk/benefit ratio. In the case of female operators using a computer in the average office, I advise that in addition to the thirty-inch precaution, as soon as a pregnancy is established the individual be relieved of computer duties until delivery. Furthermore, in those office situations involving large numbers of computers, attention should be paid to the spacing between the machines and their rows to prevent irradiation of operators in one row from the machines adjacent to them or in the row behind them. The only way to ensure a reasonably safe situation is to actually measure the field pattern and make the necessary adjustments.

A special caution is required concerning the use of computers in schools. In addition to its obvious advantages in business and personal life, the personal computer is a superb teaching device. The present tendency is to introduce schoolchildren to computers at earlier and earlier grade levels. The same precautions as noted above should very definitely be used in this situation. In addition, in view of the present uncertainties about dose-rate effects, I consider it prudent to limit computer instruction to as short a time as possible each day.

It has been my observation that many schools are using computers donated to them by people who have upgraded their own machines and given their older models to a local school. Many of these machines, while perfectly usable, were manufactured prior to 1983 and so give off substantially higher levels of radiation. Considering the fact that schoolchildren—particularly those in the elementary grades—are still growing and have a constant level of cell multiplication within their bodies, it would seem to be a wise precaution to take practical steps to mitigate the field exposures during computer instruction. Unfortunately, because there has been no official recognition of the general risk associated with computer use, there are no rules, or even recommendations, for their use in schools.

Fluorescent Lights

Fluorescent lights are much more economical than the older, incandescent bulbs. They produce many more units of light (lumens) for much longer periods of time, at considerably lower electric-power consumption. As a result, these lights have largely replaced the incandescent bulbs in most offices, schools, and public buildings.

Initially, concerns were voiced over possible biological effects from the major differences in spectral distribution of the light from fluorescent units compared to incandescent bulbs. While the light from each type *looks* almost identical, it is not. Most of the light output from a fluorescent bulb is in very small areas of the visible spectrum, while the incandescent is much more of a broadband emitter in the visible range. The incandescent light is, therefore, more nearly like natural sunlight.

However, there is a basic difference in *how* the light is produced in each case. The incandescent lamp simply causes a resistive filament to glow when current at 120 volts and very small amperage is forced through it. The fluorescent lamp has no filament. Instead, a coating of chemical on the interior of the tube is excited to glow when a high-voltage discharge is produced within the tube. This requires that the 120 volts of house current be raised to several thousand volts by means of a transformer.

The fluorescent light produces much different magnetic fields compared with the incandescent bulb. Two inches away from a 60-watt incandescent bulb, the 60-Hz field is 0.3 milligauss; at six inches, the field is 0.05 milligauss; and at one foot, the field is lost in the ambient field. A 10-watt fluorescent tube produces a field of 6 milligauss at two inches; at six inches the field is 2 milligauss; and one foot away it is 1 milligauss. The 10-watt fluorescent lamp produces a magnetic field at least twenty times greater than a 60-watt incandescent bulb. The circular fluorescent tubes of the type often installed in floor and desk lamps produce a similar field, and the user's head is often only a foot or so away. The ceiling fluorescent fixtures, with several 20-watt tubes, produce a field greater than 1 milligauss near the heads of office workers below.

Some years ago, Dr. John Ott visited my laboratory and presented his data on the behavior of children in schoolrooms that had fluorescent ceiling fixtures. While his data were impressive, they included mainly videotapes in stop motion and did not lend themselves well to quantification. Ott was convinced that behavioral disturbances in classrooms furnished with fluorescent lights were due to the abnormal light output.

Considering the average time of exposure, the magnetic fields produced by fluorescent lights can reach dangerous levels. At this time, no studies have been reported, or even started, on the bioeffects of working in an office or sitting in a schoolroom lighted with multiple banks of overhead fluorescent fixtures. From a risk/benefit point of view, the small amount of monetary savings gained by the use of these lights hardly justifies the potential risks.

Electric Clocks

An electric clock plugged into a wall socket produces an amazingly high magnetic field because of the small electric motor that runs it. A small bedside alarm clock of this type will produce a field of as much as 5 to 10 milligauss two feet away. If the bedside table is placed close to the bed, so that the sleeper's head is within this range, the dose rate is considerable for the average eight hours per night. Battery-operated clocks have a negligible field, and I recommend their use as a substitute.

Hair Dryers

Because hair dryers require high currents to produce heat, they generate very substantial magnetic fields. A 1200-watt model will produce 50 milligauss at six inches and 10 milligauss at eighteen inches from the front end of the dryer. However, the dose rate enters the picture. For the average user, the time of exposure is similar to that with the electric razor.

The hand-held hair dryer, however, appears to be a hazard for hairdressers who use them repeatedly each workday. Most operators tend to have a habitual way of using the dryers, and so they repeatedly expose certain portions of their bodies (such as the breast). We have only anecdotal evidence that cancer of the female breast is more common among beauticians than among the general public. No formal epidemiological studies have been done, and when the question of bioeffects is raised, the operators' use of many different chemicals in their trade is generally blamed.

Electric Heaters

Baseboard electric heaters are often installed in homes and offices as either the primary heat source or as a backup for some other system. A four-foot-long baseboard heater produces 23 milligauss at a distance of six inches; at one foot, the field is 8 milligauss; at two feet, it is about 3 milligauss; and it falls to 1 milligauss at three feet. In most instances, prolonged exposure occurs at this latter level, and these devices are probably not hazardous under normal circumstances. However, some caution should be observed in locating infants' cribs close to a baseboard heater. The small portable electric heaters that plug into a wall socket produce about the same field intensities, but they are even more dangerous because they may be placed much closer to the body.

In recent years, electric heat cables have been installed in ceilings as a more efficient means of heating an entire room, rather than just the periphery, as in the case of baseboard heaters. In a recent study, Dr. Nancy Wertheimer found that this method produces an average field of about 10 milligauss throughout the entire room. She also determined that the miscarriage rate among pregnant women living in such homes is higher than that of other women. This study, which has not yet been published, is significant, because in this type of installation body heating cannot in itself be a causative factor for the increased incidence of miscarriages. (Increased body heat has been associated with fetal loss, and this possibility had been raised in the case of Wertheimer's electric-blanket study.)

In order for heat to be produced, energy must be expended; in the case of electricity, this requires electric current that is roughly proportional to the heat output. While the magnetic-field strength falls off rapidly with distance, any device that electrically produces heat will have a major field strength at short distances from the device. The electric stove, for example, produces 50 milligauss at eighteen inches above a twelve-inch burner. The field strength drops off in the normal fashion, and the usual exposure to this field is not constant or long term.

Microwave Ovens

The microwave oven is another common item in American homes. As with the personal computer, the original models of these ovens were permitted to leak radiation at a higher level than models produced since 1983. However, even in the case of the latest models, leakage of one mW/cm^2 is still permitted. The regulations require only that the oven meet the required level of leakage when produced at the factory; what happens afterward is up to the consumer.

Damage to the gasketing material around the door will markedly increase the level of microwave radiation that leaks from the device when it is in use. Therefore, one should have a microwave oven checked at least once a year. If the door gasket is damaged, the device should not be used until it has been repaired and then rechecked. There are many microwave-leakage detectors on the market, but at this time there are no regulations for these devices. As a result, some may work reasonably well, while others may be quite useless. The best thing to do is to have a qualified repairman check out the oven each year. In general, these people use meters that are reasonably well calibrated.

The problem remains, however, that the "safe" level of microwave exposure has yet to be determined with any accuracy. Earlier in this book, I listed a consecutive series of studies done over the past twenty years showing that the exposure levels required to produce biological effects have steadily dropped, until now they are far below the thermal level. We still do not have any idea of the safe level for continuous exposure. Dr. Arthur Guy's study, described in chapter 8, indicates that the level must be less than 0.5 milliwatt per square centimeter. We have no idea of the time/dose relationship for intermittent exposure from a microwave oven. At present, I would say that, provided that the oven is of quality manufacture and is checked yearly by a qualified serviceman, it is probably safe for intermittent residential use. But because you shouldn't stand in front of it for the entire time that it is operating, you should give careful consideration to where you locate your microwave.

Personal Radio Transmitters

The use of personal radio-transmitting devices has risen sharply over the past decade. Formerly, these devices were licensed and controlled, and they were used primarily by radio amateurs and various public-service agencies (such as the police and fire departments). Such devices have now expanded into a host of new, unlicensed markets. These include citizens-band radios (CBs), cordless telephones, cellular telephones, home and business security systems, radio-controlled toys, and more. This development has markedly increased the number of persons exposed to significant levels of radio frequencies (RF).

The antennas of most CB radios and cordless telephones are only an inch or two away from the side of the user's head. While the amount of radiated power is limited for these devices, the restrictions are designed to prevent interference with other radio communications, *not* to protect the user. The only way to communicate with any of these devices is through generating an electromagnetic field. The same field is present in your brain when you press the "transmit" button on a hand-held device.

No studies have been done on the incidence of diseases, such as brain tumors, in people who make use of these transmitters. However, Dr. Samuel Milham of the Washington State Department of Health has reported that the incidence of leukemia is significantly higher among amateur radio operators than among the general population. Until appropriate epidemiological studies are done, I advise people to use these devices only when necessary, and then for the shortest periods of time possible.

How to Make Your Own Field Measurements

Dr. Edward Long, a general practitioner in Humboldt, Kansas, has discovered an easy, inexpensive, and surprisingly accurate method for measuring the electric fields generated by many devices. Long found that small battery-operated AM radios are very sensitive to these fields. Even though they do not respond to magnetic fields, the electric field is a reasonable index for the magnetic field itself.

To check the level of radiation from your TV set, for example, simply turn on an AM radio, tune it to a spot on the dial where you cannot hear a station, and turn the volume up to maximum. Hold the radio about a foot away from the front of the TV and switch the TV on. You will be surprised by the level of "noise" from the AM radio. If you then move the radio away from the TV, you will reach a distance at which the noise disappears. This is *approximately* the 1-milligauss level. You will have to rotate the radio at each spot you are measuring in order to get the maximum noise, since the antenna is directional. This method can be used with a number of other devices that give off RF, such as computers, stereos, and so on. It will *not* work with devices that give off 60-Hz only, such as electric stoves and hair dryers.

Most of the accurate monitoring instruments for everything from the 60-Hz power frequency to the high-frequency microwave ranges are expensive to purchase or rent. Recently, a few relatively inexpensive but sensitive devices for measuring 60-Hz fields only have become available. The reference list for this chapter contains the names and addresses of some suppliers.

In this section, I have tried to provide only general guidelines. The standards I apply are mine alone and must be considered provisional, in that they are based upon my analysis of the present data and may require revision as more information becomes available. In general, I advise people to think about the devices they currently use and to keep the time/dose relationship and the 1-milligauss recommendation in mind. It is also desirable to consider the safety factor when contemplating the purchase of a device that gives off electromagnetic radiation. Before buying it, ask yourself, "Do I really need this?" and "How long will I be exposed to the field it produces?"

AMBIENT ENVIRONMENTAL FIELDS

TV and FM/AM Radio

In the preceding section, we discussed the 60-Hz ambient field in the home that is produced by the electric-power distribution system. If you can receive television and radio programs in your home, fields of those radio frequencies are also present. While they may be exceedingly small compared to the 60-Hz field, we really do not know what the safe level of RF field strength is for continuous exposure.

This may appear to be absurd. After all, such communication is the norm in developed countries. We have been using it for a long time, you may think, and nothing has happened. Neither is exactly true. The extensive use of radio communication has occurred only in the past few decades. I am old enough to remember the "old days" of radio, in the 1920s and early 1930s, when the usual receiver was a crystal set and when only a handful of AM stations could be heard. The remarkable expansion of this modality has occurred just since World War II, with the advent of commercial television and FM radio transmissions.

The belief that nothing untoward has happened is based upon the concept that if these fields were truly harmful, they would produce a unique and very evident disease. For instance, if everyone's hair had turned green soon after the start of FM transmissions, the link would be obvious. The problem, as we have already noted, is that the disease patterns resulting from exposure to man-made electromagnetic fields are, in general, not new. Instead, for the most part they are increases in preexisting disease types, which may be attributed to many other causative factors (such as toxic chemicals). Until a sufficient number of controlled epidemiological studies have been done, we simply will not know the true extent of hazard.

In chapter 8, I mentioned the work done by Dr. William Morton in the early 1970s on the relationship between FM radiation and leukemia in Portland, Oregon. He reported that those neighborhoods with the highest levels of FM radiation also had the highest rates of leukemia. When Morton did his study, the results were considered so outlandish that the EPA, which had funded the study, ignored it. However, those results were confirmed by the 1986 study conducted by doctors Bruce Anderson and Alden Henderson in Honolulu.

Unfortunately, except for the FCC's establishment of new rules for field strength in the areas around the bases of new broadcast towers, little attention has been paid to either of these two studies. No federal guidelines have been provided for the general population, and the EPA recently announced that it will not issue any in the immediate future. It is, therefore, up to individuals to determine whether they, or their children, are at possible risk from such radiation in their own homes or schools.

Lacking the necessary hard data, the best you can do is determine how close your home is to any radiating source. The obvious question is, How close is too close? Unfortunately, the lack of data makes it impossible to provide a firm answer.

When dealing with higher-power sources, such as commercial TV or FM stations, I believe that to be reasonably safe you should be about a half mile away unless there is a major obstacle, such as a hill, between you and the tower. This is, at best, a very rough guess, based upon the preliminary data from the two studies above. I must emphasize that people living closer than this to a tower *may not* be at risk. However, it appears obvious that if your home is located within 100 feet of such a tower, a real risk is probably present, particularly for children.

In order to make an informed evaluation of the potential risk, you must know what to look for. Because it is the nature of TV, FM, and microwave to radiate as "line-of-sight" waves in straight lines from the transmitter, the tendency is to place the towers on the highest ground available, thus expanding the area within which the signals can be received. It is, therefore, common to see a number of towers clustered together on hilltops around cities or on the tops of the highest buildings in cities. These include TV, FM, and communications towers for local police and fire departments, public-service agencies, paging services, and so on. Even though each antenna is transmitting at a different frequency, the total field that results is the additive one from all of the transmitters, and the local field will be very high.

Some years ago, the EPA surveyed the entire United States and identified about 200 "hot spots" of high field strength caused by such clustering of antennas. Because of the lack of regulations, it is not unusual to find residential developments located in and around such hot spots.

AM radio stations, on the other hand, are most often located in valleys, because their signals are of a different type and do not operate as line-of-sight waves. Because the majority of AM stations were built some years ago, housing developments, schools, and public buildings have frequently been built in close proximity to them.

While the antennas and the towers come in a variety of styles, the fields given off by TV, FM and AM radio, and communication stations are quite similar, in that they all radiate in a 360-degree pattern from the tower site.

AM radio stations have several towers, located about 100 feet apart. Their signals are of a low enough frequency that they will follow the contours of the ground. Therefore, hills are no obstacle to receiving their signals. The AM radio waves that extend upward at an angle are reflected back off the ionosphere at the same angle; in this way, the signal may be heard several thousand miles away (this is the technique used for shortwave radio transmission). TV and FM waves that go up are not reflected off of the ionosphere but

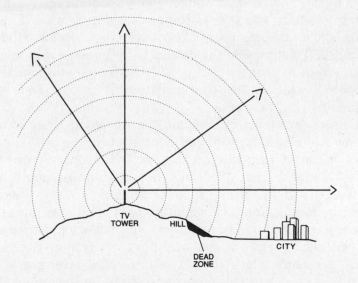

FIGURE 12-3. *Pattern of radiation from a TV or FM tower transmitter. The field extends out in all directions in straight lines. In order for the signal to be received, the receiver must have a clear line of sight to the transmitter. The signal may be blocked by a hill or tall building, and a "dead zone" will be formed behind the obstruction in which the signal cannot be received.*

simply keep going out into space. Very rarely, a solar storm will increase the energy in the ionosphere, and the high-frequency waves will be reflected back to the Earth, also to spots many thousand of miles away from where the signal originated.

Microwaves

Following World War II, most of the original long-distance telephone cables were replaced with microwave links. Fortunately, this use is now being phased out in favor of laser light along buried fiber-optic cables. Not only is this new technology more efficient, but it also appears quite safe because little electromagnetic energy is given off by such cables. However, the changeover has proceeded slowly, and only a small portion of the national long-distance microwave network has been replaced. At the same time, microwave usage by various public-service agencies, utilities, and commercial establishments is growing.

When questions of safety are raised over an existing or proposed microwave installation, the public is assured that the radiation is given off as a narrow or "pencil-like" beam directly toward the receiving antenna, and that nothing else is irradiated. The fact is that this is what the engineers *wish* it were like, because that would be the most efficient use of the power. In actuality, every microwave antenna emits a primary beam with accompanying, unavoidable "side lobes," which spread the radiation around a 180-degree arc from the dish.

Microwave antennas also come in a variety of shapes and sizes, and many are often clustered on a single tower.

When considering the possible hazards from microwaves and radio-frequency sources of all of types, take into account not only your distance away from any such installation, but also the fact that any specific location may be radiated by several transmitters. It is therefore necessary to measure the field from each source and add them together to arrive at the total field strength at your location. It is my personal opinion that any total field strength above 0.1 milliwatts per square centimeter is likely to be hazardous for residential exposure.

The only way to determine whether your home falls within this range is to have measurements taken or to try to do it yourself with rented microwave and RF monitors. Even then, lacking any established federal guidelines, it would be practically impossible to have

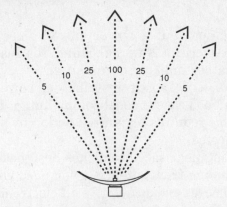

MICROWAVE DISH

FIGURE 12-4. *Pattern of radiation from a typical microwave dish antenna. The main beam is here considered to contain 100 percent of the power. There is comparably lower-strength radiation coming out from the dish in a fan pattern. The numbers refer to the percentage of total power contained in these "side lobes."*

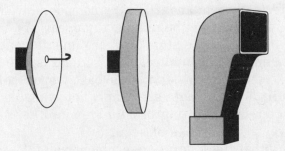

FIGURE 12-5. *Typical types of microwave antennas. At left is the standard dish type, which may be made of a solid metal or of meshwork. The signal is sent through a wave guide to the feed horn, which is located above the center and faces the dish surface. The microwave radiation is given off from this feed horn toward the dish, which reflects it in the desired direction.*

The center illustration is of a drum-type antenna, which is the same as a dish except that the front is covered with a material that is transparent to the microwave signal.

At right is a horn-type antenna. The signal comes up to the horn via a wave guide. It is then turned, and it comes out horizontally from the opening of the horn.

All three types produce the same radiation pattern, as shown in Figure 12-4.

an existing installation shut down or moved unless the measured
levels exceed the American National Standards Institute guideline
of 5 mW/cm², a rather unlikely occurrence.

In the case of a *proposed* installation, however, it is often
possible to make a good case to local zoning authorities on potential
health hazards. This requires organized action by the citizens who
are potentially affected.

If you are living near satellite dishes designed only to *receive*
television programming from satellites, you need not be concerned;
these are passive devices and do not give off electromagnetic energy.
Satellite-*transmitting* dishes, however—which are from fifty to
eighty feet in diameter—radiate strong fields.

THE LONG-TERM RADIO FREQUENCY/ MICROWAVE SOLUTION

The problem of RF contamination of the environment
must be addressed at the federal level. Some states and municipali-
ties have already begun to set local standards, based on the potential
for health effects. Since standards vary considerably from state to
state, a patchwork effect is beginning to appear. As a result, on a
federal level many of these standards are unenforceable. In order
for us to have enforceable and realistic standards, the federal government must set uniform standards based upon solid scientific
data. Unfortunately, such data are simply not available at this time.
Therefore, I oppose the setting of *any* standard, on the basis that
it would convey a false sense of security.

Assuming that the federal government will take up this problem in a realistic fashion, what will happen if the hazard level falls
below that which is already present in most metropolitan areas? Will
we have to turn off our TVs? Not at all. Once the hazard level has
been determined, the engineering solutions are not difficult.

At present, the FCC limits the radiated-signal power from
every TV, FM, and AM station to prevent interference with other
stations. Once the need to reduce the total ambient levels has been
determined, average signal power may be reduced to meet that
requirement. Additional steps may include limiting times of operations and using lower-power repeater stations to extend areas of
coverage.

The clustering of TV towers can be replaced with direct satellite broadcast (DSB). In this technique, the programming is trans-

mitted by satellite to a wide geographic area, with very low average power levels at the ground surface. At present, this method is in use in many rural areas where tower installations would not be economical.

There are other steps that can be taken, and there is no need to turn off our public communication systems in order to make them safe.

THE LONG-TERM ELECTRIC-POWER SOLUTION

In the case of transmission lines and their auxiliaries, such as substations and transformers, we now have the action level of a 3-milligauss magnetic field upon which to base a risk assessment.

If you are living in close proximity to an established transmission line and measurements inside your home average 10 milligauss, what can you do? If you have concluded through a risk/benefit determination that you and your family are at risk, the answer is simple but unpleasant: your only option is to move. As with microwave and radio-transmitter installations, legal actions against established transmission lines are costly, prolonged, and fruitless. However, concerted action may often succeed in preventing the installation of a proposed line. Nevertheless, be aware that such an effort is still expensive in time, effort, and money.

Utility companies will defend proposed lines on two grounds. First, they claim that there is no compelling evidence for health hazards from transmission lines. Second, they say there is a pressing need for the line to supply more electrical power. They will claim that if it is not built, the economy will suffer, and insufficient power will lead to brownouts or blackouts. Both claims may be easily refuted, the first on the basis of the data contained in chapter 8, and the second on the basis of the long-term solution to the problem.

Do we really need more electric power? While utility companies are regulated by each state, they are still businesses, and they are continually attempting to sell more power and justify the building of more generating plants and transmission lines. The record shows that they have consistently overestimated the future requirements for electrical power.

Robert C. Marley, special assistant to the deputy assistant secretary for conservation in the U.S. Department of Energy, has studied trends in the industrial use of energy since 1973. In 1984, he reported,

During the ten years following the 1973 Arab oil embargo, significant changes took place in U.S. energy consumption. After decades of steady growth, annual demand for energy leveled off and began to decline. Total energy consumption in 1983 was less than that in 1973, despite economic growth over this same period averaging 2.5 percent per year.

Prior to 1973, the utilities were predicting a 30 percent increase in demand by 1983. This obviously did not occur. Nevertheless, the economy continued to grow. Marley's point was that the growing availability of cheap energy in 1984 could again lead to increased consumption, which would actually be unnecessary and inefficient.

Two years later, Marley's concerns became fact. The utilities again began issuing forecasts of impending doom if new construction was not begun immediately. In addressing this issue, Dr. Amory Lovins of the Rocky Mountain Institute in Old Snowmass, Colorado, showed that increasing the efficiency of utilization of electrical power could easily produce the equivalent of the increased generating capacity, and at a much lower cost.

To put the situation more bluntly, in a briefing on nuclear power issued in 1988 the Union of Concerned Scientists stated,

Today, the U.S. is overflowing with excess electricity. In 1985, we had a reserve of nearly 52 percent. That means the nation had much more electricity than it needed to meet its peak summer demand, as well as an adequate margin for emergencies. Utilities had many more large power plants than necessary to provide reliable electric service.

It is obvious that at this time, we really do not need more generating plants and transmission lines. All of the above analyses were concerned only with the *efficiency* of operation of the U.S. energy system. None took into account any possible health risks to the general population from 60-Hz power systems. When this factor is included in the analysis, the conclusion becomes even more obvious, and the requirement for conservation and increased efficiency even more pressing.

I am not advocating the dismantling of the electrical-power supply system and its replacement with kerosene lamps. What I do advocate is a change in the basic structure of the entire system. The concept of a single large generating plant supplying a wide area with electrical power via long transmission lines (the concentrated system) arose when electrical power first began to be generated. At that time, many small, independent systems were set up, each supplying relatively small areas. The motivation for the concentrated system was the realization that it was possible to increase company profits by acquiring these smaller systems and supplying their combined area from a single large generating plant. Since then, the concentrated-system technique has dominated the entire industry.

When tied together into a nationwide transmission power grid, the concentrated system has some important defects. First, it is a fragile system: failure of a single component can cause the crash of large portions of the system, with subsequent blackouts lasting for several days. Second, as the trend toward greater and greater concentration of generating plants grows, the need for super-voltage transmission lines increases (765-kV lines are already in use, and 1-million-volt lines are planned). The fields produced by such lines are proportionally larger in extent and higher in strength, and their construction will result in the exposure of greater numbers of the population to potentially harmful levels of electromagnetic radiation.

Both of these defects could be obviated by a restructuring of the system into a dispersed type, in which single, large generating plants are replaced by multiple smaller ones. The smaller plants could be located to take advantage of multiple generating sources, such as water, wind, and the sun, in addition to the standard fuels. The transmission lines from each small plant would supply a smaller area and would require much lower voltages and currents. Each generating plant and supplied area would be tied to its neighbors in such a fashion that power could be supplied from neighboring sources in the event of a shortage or outage.

In addition to producing much less hazardous fields, a dispersed system would be economically competitive with the concentrated system and at the same time would be less susceptible to large-scale blackouts. The time required to accomplish the conversion from the concentrated to the dispersed systems would be considerable, and such a change would be fiercely contested by the utilities themselves. Nevertheless, in consideration of the percentage of the population that may be already at risk from 60-Hz electric-power systems, the problem must be addressed as a national objective.

If advances in photoelectric technology continue, the above steps may turn out to be unnecessary. In photoelectric technology, sunlight is converted directly into electric power by collectors. In appropriate locations with abundant sunlight, efficient collectors could theoretically supply all the power for a single residence. At present the efficiency of these devices has risen to slightly more than 10 percent. If it can be increased to 15 percent, the photoelectric system will become economically competitive with commercial power systems.

RISK/BENEFIT ANALYSIS: INDUSTRY'S VIEW AND CITIZEN ACTION

The idea that we can live in a totally risk-free world is not only unrealistic but unattainable as well. There are very few things in our society that are solely beneficial or solely devoid of benefit. Everything is some shade of gray, and the best that we can do is minimize our personal risks consistent with benefits. The sticking point comes when risk/benefit decisions have to be made on community or national levels. These decisions are generally not directly in the hands of the citizens who would be exposed to the risks; instead, they fall to various governmental bodies.

When installations are proposed by private industry (utilities, manufacturers, biotechnology companies, and so on), the question of risks versus benefits is raised early on, often by the industry itself. The usual statement issued is that the benefits far outweigh the "minimal" risks involved. Since risk/benefit analyses are made on the basis of scientific evidence, scientists inevitably enter the process.

There are two types of scientists who become involved in the issue of electromagnetic-field health risks. First are the experts on the biological effects of electromagnetic fields, who often take diametrically opposite views. Some emphatically insist that no harmful effects can occur, while others are equally adamant that real hazards already exist. Less common, but increasing in number, are the experts on risk/benefit analysis itself. These people are not necessarily directly involved with the specific subject, but they consider themselves able to evaluate the risk/benefit ratio for *any* question. They do this on the basis of a statistical, computerized analysis of the scientific literature, arriving at numerical data that compare the risk in question with other risk-producing situations. For example, these scientists can calculate the number of expected deaths from cancer in a population exposed to a 60-Hz electric power-transmission line, compare that number with the expected number of deaths in the same population from cigarette smoking, and declare that the transmission line poses less hazard. While all of this may *look* very scientific, I believe it is nonsense. The basic defect in this type of analysis is that the present scientific data are accepted as the final word, never to be revised or updated.

I have yet to see a scientific question over which there was no difference of opinion among reputable scientists. Such differences may be honest, motivated by some uncertainty about the data or by a valid but different outlook. When this occurs, the actual decision-making authorities have a difficult time. There is a well-known story about a Senate hearing during which such a situation arose. A senator was overheard to whisper to a colleague, "What am I supposed to do, count the Nobel Prize winners on each side and say that the one with the most wins?"

Unfortunately, expressed differences of scientific opinion are not always so noble. Scientists are, after all, human beings, and some can be tempted to shade their opinions for certain considerations. It is quite possible to slant a presentation by ignoring positive data indicating risk, or to require of such data a much higher standard of scientific rigor than for negative data. The usual recommendation from this type of expert consultant is that some risk may be present, but that it requires more study in order to be clarified and, in any event, is not significant enough to prevent an installation from being constructed. The citizen groups in opposition to the planned construction generally lack the resources to hire scientific experts who share *their* views.

Having been involved in such situations more times than I would have liked, I am of the opinion that legal steps should be taken to require the industry or governmental agency involved in such a situation to correct the inequality by providing the citizens in opposition with the same amount of dollars that the industry or agency itself spends on expert testimony. It seems very unlikely that this will ever come about.

The answer to a valid risk/benefit determination appears to me to be quite clear. All of the data needed for making a decision should be made available to everyone concerned. This seems simple enough, but in our technological world it often means the expenditure of considerable sums of money. If the data are inadequate or incomplete, a major benefit over any possible risk should be demonstrated before a conclusion is made that the benefit justifies the risk. If the data are adequate, the decision should be based upon the open and unbiased evaluation of these data—something that is more easily said than done. Of equal importance to adequate data is the principle that the only persons who can make the final decision as to risk versus benefits are those persons who would be at risk.

If citizens believe they will be placed at risk by an installation, the first thing they should do is organize. A citizens' group that is well organized has two effective weapons. The first is the press and media. All that is needed is to make the story of the opposition newsworthy, and to keep it so. Representatives of news organizations have the resources to make the citizens' position much more evident and to reinforce it by interviewing experts who hold supportive scientific opinions. The citizens' second weapon is the ballot box at the next election.

Some corrective measures are urgently needed in this entire process, not only in regard to electromagnetic-field–producing installations, but also to the entire range of technological innovations that impinge on the general environment. Until appropriate measures are taken by the responsible federal agencies, it is up to individual citizens to make their own risk/benefit analyses, and to take those steps that they themselves feel are prudent to protect their health and the health of their children.

Epilogue

I hope that this book has revealed at least a few of the defects in the chemical-mechanistic theory as well as the new vision of scientific vitalism that has sprung from the latest scientific revolution. This new scientific paradigm is not vitalistic in the old sense of a mysterious "life force" that would remain forever beyond human knowledge. Ironically, to many it may appear to be an even more mechanistic view than the current doctrine.

But while the new paradigm is an extension of the old one, it is an extension in a *new direction*—a direction that has taken us deeper into the mystery of life, to the level of the energetic-control systems that make the living organism greater than the sum of its parts. While the need for such systems has long been recognized, the new scientific revolution has identified the energies involved as electrical currents and magnetic fields.

This knowledge has given us a better understanding of how the human body uses these energetic systems to heal itself and to regulate its activities. It also has enabled us to reevaluate previously discarded medical therapies and to explore new ones based upon influencing these same systems. This developing knowledge will enrich and supplement, but not replace, present medical techniques.

I believe that the mechanistic idea has done almost irreparable harm to medicine. First, in its proponents' arrogance that it alone was the truth, it gave rise to the ultimate technological medicine, in which physicians lost sight of life in the complexities of molecular biology. Second, because all living things, including humans, were seen as simply chance occurrences, a patient's status became nothing more than that of a machine in need of repair. Medical ethics, based on the uniqueness and sanctity of life, became a secondary concern.

Finally, the chemical-mechanistic doctrine has led to a complex web of interrelationships among physicians, pharmaceutical and equipment manufacturers, and regulatory agencies. Driven by the profit motive, this has resulted in an out-of-control upward spiral of medical costs. For obvious reasons, the therapeutic concepts arising from the new paradigm are being vigorously contested.

These new concepts of life and energy bring back into medicine a humility before the miracle we call life. The *art* of medicine becomes a practice composed of the life energies of the physician, the patient, and the Earth.

The new scientific paradigm of life, energy, and medicine has also led us to reconsider many of our technological "advances," which have separated us more and more from the unique electromagnetic environment that has been home to life since its beginning. The evidence is clear that our unrestricted use of electromagnetic energy has produced a global environment that is more and more hazardous to life. The time within which it is possible to take remedial action without sacrificing the real advantages of this technology is shrinking.

Unfortunately, there are strong forces that, for the sake of perceived immediate gain, prefer to believe that unlimited time remains. Foremost among these are the military establishments of all nations. Present strategic doctrine rests on the unlimited, and expanding, use of electromagnetic energy. Without this capability, sophisticated weapons systems are useless. As a result, any attempts to acquaint the general public with the potential hazards of electromagnetic fields are viewed by such forces as inimical to state security and so are ruthlessly suppressed.

An informed public is the only defense against the forces of the military and medical establishments that seek to prevent further development of the new concepts of life energy and medicine. It was for this reason that I undertook the task of writing this book.

The Hidden Hand on the Switch: Military Uses of the Electromagnetic Spectrum

The military group provides powerful incentives for releasing forbidden impulses, inducing the soldier to try out formerly inhibited acts which he originally regarded as morally repugnant.

I. L. JANIS,
as quoted in *The Oxford Companion to the Mind*,
edited by Richard L. Gregory

*T*he current military strategy of the United States does not rest on nuclear weapons or elite strike forces, but is instead based on a doctrine known as C³I (for command, control, communications, and intelligence). Through C³I, the relative strength of our forces against those of the Soviet Union is constantly assessed, and the intent of the Soviets to deploy and use their forces against us is determined. C³I provides intelligence to accomplish this task, as well as the means to communicate with and control our forces and to command them to counter perceived or actual threats. The overall intent of this military doctrine is to know, instantaneously and at all times, the exact status of our forces and those of a particular enemy. Weapons systems are simply the instruments that are used to deter or respond to an attack.

Our total military system has become an analog of a living organism, constantly sensing its environment, integrating information, and reaching decisions, and then acting on those decisions by using the appropriate weapons systems. The "central nervous system" of our nation's global military organism is based on information-carrying electrical impulses that are transmitted by electromagnetic fields. Its sensory organs are microwave scanners, satellites, and sophisticated devices designed to listen to an enemy's radio transmissions. Instead of nerves, it uses radio communications with frequencies ranging from ELF to superhigh microwaves. The muscles of the military organism range from ground troops to nuclear missile systems. There are a number of "brains" in the organism, located both in the continental United States and in various overseas locations. The organism is capable of operating on its own, with only the theoretical restraint of approval from the White House. Every aspect of C^3I depends upon the unrestricted use of all frequencies of the electromagnetic spectrum at unlimited power densities.

HISTORICAL DEVELOPMENT

This military doctrine slowly evolved following the end of World War II and was shaped by two factors. The first was the practical experience of using electromagnetic fields for communications and sensing (primarily radar) during the Korean War. The second factor was the later availability of transistorized equipment and the development of exotic electronic sensing and communication systems, which occurred during the Vietnam conflict.

Vietnam was the proving ground for the basic concepts of C^3I, and it has been characterized as the first all-electronic war. A number of advanced technologies were tested—for example, long-range reconnaissance patrols operating far behind Vietcong lines were equipped with solar-powered, high-frequency radios. These devices enabled members of a patrol to communicate with one another via military satellites 200 miles up in space and to be in instantaneous communication with the White House. Following the Vietnam war, the doctrine of C^3I matured into its present global scale based on maximum utilization of electromagnetic energy.

The only restriction ever placed on this military doctrine since its inception in the 1940s was derived from the period of time immediately following World War II and during the beginning of the Cold War. In the early 1950s, the Department of Defense recognized the need for some sort of "safe exposure standard" for microwaves. This led directly to the establishment of the Tri-Services Program, based at the Rome Air Development Center in Rome, New York, which was given the task of determining this standard. However, even before the Tri-Services Program began, the military eagerly adopted the concept that only thermal effects were damaging to living organisms. Based solely on calculations, the magic figure of

10 milliwatts per square centimeter was adopted by the air force as the standard for safe exposure. Subsequently, the thermal-effects concept has dominated policy decisions to the complete exclusion of nonthermal bioeffects.

While the 10 mW/cm² standard was limited to microwave frequencies, the thermal concept was extended to all other parts of the electromagnetic spectrum. Unless it heated tissue, electromagnetic radiation was thought to be harmless, so there were no limits placed on exposure to frequencies below microwave.

THE CONSPIRACY

The military organism was designed on the 10-mW standard and, once in place, it had to be defended against the possibility of nonthermal bioeffects. The recognition and validation of these effects would mean the collapse of the total organism and the death of C³I. My work on electrical control systems and the bioeffects of electromagnetic fields involved me in this controversy early in the 1970s. It quickly became apparent to me that evidence for nonthermal effects was viewed as a threat to national security. Safety was not a consideration, because the military mind-set of the time held that despite the lack of actual hostilities, we were in a state of war with the Soviet Union. It was believed that our ability to prevail in that conflict required the virtually unlimited use of electromagnetic energy for all four facets of the C³I doctrine.

This view led to the policy of denying any nonthermal effects from *any* electromagnetic usage, whether military or civilian. To accomplish this policy objective, several specific actions were taken, as follows.

Control over the scientific establishment was maintained by allocating research funds in such a way as to ensure that only "approved" projects—that is, projects that would not challenge the thermal-effects standard—would be undertaken. Further, the natural reactionary tendency of science was capitalized upon by enlisting the support of prominent members of the engineering and biological professions. In some instances, scientists were told that nonthermal effects *did* occur, but that national security objectives required that they be exceptionally well established before they became public knowledge. Many scientists' goals were subverted by unlimited grant funding from the military and by easy access to the scientific literature.

The formal scientific establishments of the United States

were mobilized. When serious challenges to the thermal-effects standard were raised publicly, eminent scientific boards, associations, or foundations were provided with lucrative "contracts" to evaluate the state of knowledge of bioeffects of electromagnetic fields. These investigations resulted in the production of voluminous "reports."

All of these reports shared certain characteristics. Scientific data indicating nonthermal bioeffects were either ignored or subjected to extensive and destructive review. Those examined were required to have much higher standards of possible validity than reports indicating no such bioeffects. Scientists who reported the existence of nonthermal bioeffects were ridiculed and were portrayed as being outside of the mainstream. Actual disinformation was utilized to create a false impression: for example, while a statement such as "There is no evidence for any effects of pulsed magnetic fields on humans" would have been *literally* true, it would have ignored the many reports of such effects on laboratory animals and the fact that no actual tests had been conducted on humans. It was common practice to include an "executive summary" with the massive report. These summaries never reflected the data that were actually hidden in the full report.

A group of "manufactured" experts was produced to serve as spokesmen and expert witnesses. These were people with few qualifications for research in this (or any) scientific field, who were provided with large research grants and placed on many committees, boards, and international governmental commissions dealing with the bioeffects of electromagnetic energy. Superficially, they appeared to be prominent researchers, until one discovered that the actual number of scientific papers they had produced was minimal. These "experts" were, and still are, used to testify in legal proceedings dealing with civilian installations such as power lines and microwave-relay systems.

Scientists who persisted in publicly raising the issue of harmful effects from any portion of the electromagnetic spectrum were discredited, and their research grants were taken away.

Despite the application of these measures, the question of harmful effects did not go away but instead increased in intensity. The position of the government has thus been forced to change. The government's initial complete denial of any nonthermal effects was followed by acceptance of *some* nonthermal effects, although these were characterized as being unimportant and transient. At present, the official position is that while there are some nonthermal effects that may be harmful, further study is required before any sudden action is taken. These studies are going on, but all are under the aegis of either the Defense Department or the industry involved.

THE CURRENT SITUATION

The policy objective has been achieved, and the exposure of both civilians and military personnel to electromagnetic radiation continues. The military has not revealed its present "safe exposure standard," but there is ample reason to believe that it basically remains fixed at the now discredited 10 mW/cm^2 standard, or even higher in certain essential situations.

The reason for this is the long lead time in weapons-system development. For example, every one of our present operational microwave weapons systems was designed around the 10-mW standard. Operation at lower power levels would materially degrade system performance, allegedly producing a situation hazardous to national security. Deployment of powerful and exotic electromagnetic systems continues, with little, if any, consideration given to the potential impact of these systems on the health and safety of the public.

The Ground-Wave Emergency Network (GWEN)

While there are many such weapons systems, I consider the GWEN system to be a particularly good example. GWEN is a communications system currently under construction that operates in the very low frequency (VLF) range, with transmissions between 150 and 175 kHz. This VLF range was selected because its signals travel by means of ground waves—electromagnetic fields that hug the ground—rather than by radiating into the atmosphere. The signals drop off sharply with distance, and a single GWEN station transmits to a 360-degree circle radiating out from it to a distance of about 250 to 300 miles.

The GWEN system consists of approximately 300 such stations, each with a tower 300–500 feet high. The stations are spaced from 200 to 250 miles apart, so that a signal can go from coast to coast by hopping from one station to another. When the system is completed in the early 1990s, the entire civilian population of the United States will be exposed to the GWEN transmissions.

The rationale for the existence of this network is the government's concept that nuclear war is winnable if a fail-safe communications system is available for use during and following a nuclear attack. Such a system would permit the United States to order its nuclear-missile submarine fleet to launch an attack against the aggressor nation. The physical nature of nuclear war requires that this system operate by ground-wave transmission.

Electromagnetic Pulse (EMP)

One aspect of nuclear war that is not well publicized is the EMP phenomenon. An electromagnetic pulse is a very short, intense burst of electromagnetic energy that is produced by the explosion of a nuclear weapon in space. If an EMP were produced by a nuclear explosion 100 miles above Kansas City, its energy would be so intense that it would shut off all electric-power systems, destroy all computers and magnetic disks or tape records, destroy the guidance systems of missiles and the computer and communication systems of military and commercial aircraft, and shut down all radio communications—*across the entire United States.* The military organism would be decapitated. In the military scenario, the United States would then be faced with capitulation or nuclear destruction.

Theoretically, ground-wave communications would still be possible. However, the theory is tenuous. The GWEN hardware is transistor based; even if placed in "hardened" bunkers, it would be vulnerable to an EMP. In addition, the EMP would produce major ground currents in the path of the GWEN signals that could decrease their transmission capabilities. Finally, the locations of all GWEN stations are known to the Soviets and thus are vulnerable to attack.

Nonetheless, the military mind has conceived of using the GWEN network to maintain communications following such a decapitating EMP attack. This is not the place for a full argument concerning the values and options of nuclear war, but in my opinion the reason for the existence of GWEN is specious. Nuclear war is not winnable.

And the potential harm to the civilian population from the operation of GWEN has not been addressed. I am concerned not only
because of the data summarized in chapter 8, but also because of the
potential for behavioral and cognitive alterations that have been
discussed in this book. GWEN is a superb system, in combination
with cyclotron resonance, for producing behavioral alterations in the
civilian population. The average strength of the steady geomagnetic
field varies from place to place across the United States. Therefore,
if one wished to resonate a specific ion in living things in a specific
locality, one would require a specific frequency for that location. The
spacing of GWEN transmitters 200 miles apart across the United
States would allow such specific frequencies to be "tailored" to the
geomagnetic-field strength in each GWEN area. While I doubt that
this potential use has occurred to the planners of the GWEN network, or that such action could be deliberately taken by any portion
of the federal government, the mere *existence* of the GWEN system
may, at some future date, prove irresistible.

The New Killing Fields: Electromagnetic Weapons

While the military was vigorously denying the very
existence of bioeffects from electromagnetic-field exposure, such
bioeffects were actually being explored as potential weapons—
weapons with the enormous advantage of being totally silent and
imperceptible.

The EMP concept has been extended through the development
of devices that generate EMP pulses without the need for nuclear
explosions. Such devices could be deployed for use against enemy
command and control centers or against aircraft in order to produce
failure of electronic equipment. A derivative of this program is HPM
(high-power pulsed microwave), a system producing intense, extremely short pulses of microwave. Several types, ranging in frequency from 1200 MHz to 35 GHz with powers up to 1000 mega-
watts, are being tested. These are also being considered for potential
use as weapons against human beings.

A recent report derived from the testing program of the Microwave Research Department at the Walter Reed Army Institute of
Research states, "Microwave energy in the range of 1 to 5 GHz, a
militarily important range, penetrates all organ systems of the body
and thus puts all organ systems at risk." Effects on the central

nervous system are considered very important. The testing program, begun in 1986, is divided into four parts: (1) prompt debilitation effects; (2) prompt stimulation through auditory effects; (3) work interference/stoppage effects; and (4) effects on stimulus-controlled behavior. The report goes on to state, "Microwave pulses appear to couple to the central nervous system and produce stimulation similar to electrical stimulation unrelated to heat." It appears that HPM is capable of altering behavior in the same fashion as Delgado's electrical stimulation.

The production of cognitive and behavioral alterations by HPM is a sledgehammer effect in comparison to the subtle alterations produced by ELF fields. According to a 1982 air force review of biotechnology, ELF has a number of potential military uses, including "dealing with terrorist groups, crowd control, controlling breaches in security at military installations, and antipersonnel techniques in tactical warfare." The same report states, "[Electromagnetic] systems would be used to produce mild to severe physiological disruption or perceptual distortion or disorientation. They are silent, and countermeasures to them may be difficult to develop."

A new class of weapons, based on electromagnetic fields, has been added to the muscles of the military organism. The C³I doctrine is still growing and expanding. It would appear that the military may yet be able to completely control the minds of the civilian population.

I have made no attempt here to review in any detail the relationship between military considerations and the hazards of man-made electromagnetic fields. This complex and dangerous situation lies outside the scope of this book, except for an indication of how the political policies derived from it have effectively hampered the public recognition of the hazards. In my opinion, the military establishment still believes that the survival of the military organism is worth the sacrifice of the lives and health of large segments of the American population.

GLOSSARY

Alternating current (AC). An electrical current that changes its direction of flow with a certain periodicity. For example, 60-cycle AC is an electrical current that changes its direction of flow 60 times per second. Also loosely used as a synonym for any current or field that varies with time.

Blastema. The mass of primitive, embryonic cells appearing at the site of injury in an animal capable of regeneration. These cells subsequently grow and form a replica of the missing part.

Carcinogenic. A substance or force capable of causing a cancer.

Dedifferentiation. The process in which a mature, specialized cell returns to its original, embryonic, unspecialized state. During dedifferentiation the genes that code for all other cell types are made available for use by derepressing them. See also **Differentiation** and **Gene.**

Differentiation. The process in which a cell matures from a simple embryonic type to a mature, specialized type in the adult. Differentiation involves restricting, or repressing, all genes for other cell types. See also **Dedifferentiation, Gene.**

Direct current (DC). An unvarying or steady electrical current. Also used as a synonym for any unvarying or steady-state process, such as a DC magnetic field.

Electromagnetic field (EM field). A force field (see **Field**) that extends or radiates out away from any moving electrical current. This electromagnetic field has a direction of movement away from the electrical current and contains both a magnetic field and an electric field.

Electromagnetic spectrum (EM spectrum). A way of organizing electromagnetic fields on the basis of their frequency of oscillations. The nonionizing electromagnetic spectrum starts with zero (no oscillation, or DC) and extends up to visible light with trillions of oscillations per second. Frequencies of oscillation above light are considered to be ionizing and include X-rays and cosmic rays. The EM spectrum is divided into regions based on frequency and usage. See also **Extra low frequency, Very low frequency, Radio frequency,** and **Microwave.**

Extra low frequency (ELF). The portion of the electromagnetic spectrum extending from zero to 1000 cycles per second. This includes the 60-cycle power frequency in the United States, the 50-cycle European power frequency, and the U.S. Navy's submarine communication system at 45 and 75 cycles per second.

Field. The area around a source of electric or magnetic energy within which a force exists and can be measured. This is sometimes termed *radiation* in the sense that electromagnetic fields radiate out and away from the source and have characteristics of particulate radiation (see **Photon**).

Gene. A portion of DNA that determines a specific characteristic. See also **Oncogene.**

Hertz (Hz). The frequency of electromagnetic radiation in cycles per second. One Hz is one cycle per second. One thousand cycles per second is 1 kHz. One million cycles per second is 1 MHz. One billion cycles per second is 1 GHz. Hz is derived from Heinrich Hertz, who discovered electromagnetic radiation.

Ion. An atom that has lost or gained one or two electrons so that it has an electrical charge and is chemically much more active than the neutral ion, which has a balance of positive and negative charges.

Ionizing. The production of ions from neutral atoms by exposure to radiation with sufficient strength to dislodge electrons.

Magnetite. A naturally occurring mineral that is magnetic.

Melatonin. A hormone produced by the pineal gland that regulates the level of activity in the brain.

Microwave. That portion of the electromagnetic spectrum extending in frequency from 500 million cycles per second (500 MHz) up to the frequencies of visible light.

Mitosis. The process of cell division.

Necrosis. The dissolving of dead tissue or cells.

Neuroepidermal junction. A structure formed from the union of skin and nerve fibers at the site of tissue loss in animals capable of regeneration. The structure produces the specific electrical currents that bring about the subsequent regeneration.

Nonionizing radiation. That part of the electromagnetic spectrum extending from zero frequency to the frequencies of visible light. This radiation does not contain sufficient intrinsic energy to cause ionization of atoms in the body's chemicals.

Nonunion. A fracture of a bone that has failed to heal.

Oncogene. One or more genes that produce cancer. These genes are normally repressed, but a variety of factors can cause their activation, including virus infections, carcinogenic chemicals, and nonionizing and ionizing radiation.

Osteomyelitis. Infection of bone.

Perineural cell. One of several types of cells that surround nerve cells.

Photon. The theoretical particle that carries the force in the electromagnetic field.

Pineal gland. A small structure situated in the center of the head and connected to the brain that was originally the "third eye," which was located on the top of the head in primitive animals. See also **Melatonin.**

Radio frequency (RF). That portion of the electromagnetic spectrum extending from 500,000 cycles per second (500 kHz) to 500 million cycles per second (500 MHz).

Resonance. In general this refers to a circumstance in which some aspect of a force—sound waves, for example—has a physical characteristic that "matches" a characteristic of a physical structure, such as the mass of a building, in which case the sound waves will produce vibrations in the building. In the case of electromagnetic radiation, resonance implies a "match" between the wavelength of such radiation with the physical size of a structure. When resonance occurs, the power in the force is maximally transferred to the physical structure.

Semiconduction. The conduction of electrical current by the movement of electrons or by the absence of electrons (called "holes") through a crystal lattice. It is the third and most recently discovered method of electrical conduction. The other methods are metallic conduction, which works by means of electrons traveling along a wire, and ionic conduction, which works by movement of charged atoms (ions) in a solution. Semiconductors conduct less current than metals but are far more versatile than either of the other types of conduction. Thus they are the basic materials of the transistors and integrated circuits used in most electronic devices today.

Spectrum. A way to organize a mass of information according to some shared characteristic—for example, a spectrum of neurological diseases. See also **Electromagnetic spectrum.**

Time varying. Changing with time, such as a time-varying electromagnetic field in which the rate of change is called the frequency and is expressed as the number of oscillations per unit of time.

Very low frequency (VLF). That portion of the electromagnetic spectrum extending from 1000 cycles per second (1 kHz) to 500,000 cycles per second (500 kHz).

RESOURCES

CHAPTER ONE
THE HISTORY OF LIFE, ENERGY, AND MEDICINE

American Physiological Society. *Neurophysiology. Vol. 1 of The Handbook of Physiology.* 1959. Chapter 1, "The Historical Development of Neurophysiology," by Mary A. Brazier, is one of the best and most literate accounts of this subject. This work may be found in most medical school libraries.

Brotherstone, Gordon. *Images of the New World.* London: Thames & Hudson, 1979. A collection of actual writings of Amerindians describing aspects of their culture, including their medical and healing concepts.

Clendening, Logan. *Source Book of Medical History.* New York: Dover Publications, 1942.

Collier, Michael. *Introduction to Grand Canyon Geology.* Grand Canyon, Ariz.: Grand Canyon Natural History Association, 1980. This has little to do with medicine *per se*, but it is an excellent, poetically written history of geology and evolution that relates to scientific attitudes.

Dibner, Bern. *Luigi Galvani.* Norwalk, Conn.: Burndy Library, 1971. A brief biography and discussion of Galvani's observations with several illustrations by Galvani that were not previously published.

Grossinger, Richard. *Planet Medicine.* Berkeley, Calif.: North Atlantic Books, 1987. An encyclopedic review of medicine from prehistoric shamanism to current medical practice.

Meyer, Herbert. *A History of Electricity and Magnetism.* Norwalk, Conn.: Burndy Library, 1972.

Author's note: There are conflicting statements about Paracelsus. Most authorities agree that he was born in Maria-Einsiedeln, now in Switzerland. However, the ancestral residence of his family was Hohenheim castle, located in Plinnigen, a small village near Stuttgart, Germany. While in Germany in 1951, I visited castle Hohenheim and found that the local population claimed that "Dr. Faustus" had been born there and had often visited between his wanderings. The castle was dark, cold, damp, and quite sinister—a fit setting for Dr. Faustus.

A good general review of Paracelsus' life may be found in the *Encyclopaedia Brittanica*, 15th ed. It was written by John G. Hargrave, who also authored the book *The Life of Paracelsus*. The quote "natural extrasensory perceptionist" was taken from the *Brittanica* article. Excellent sourcebooks on Paracelsus are *The Prophecies of Paracelsus* and *The Life and Teachings of Paracelsus*, by Franz Hartmann, M.D., published in 1973 by Rudolf Steiner Publications (5 Garber Hill, Bauvelt, New York).

A somewhat similar circumstance pertains to Mesmer. Most articles list his place of death as Meersburg, a small town that the world seems to have forgotten, located on the north shore of Lake Konstanz in Germany. I also visited Meersburg and found Mesmer's house; a plaque on the door identified the house as his *birth* place.

CHAPTER TWO
THE NEW SCIENTIFIC REVOLUTION: THE ELECTRICAL CONNECTION

Bailer, J. C., and Smith, E. M. "Progress against Cancer?" *New England Journal of Medicine* 314 (1986): 1226.

Becker, R. O. *IRE Transactions on Biomedical Electronics* 7 (1960): 202. A description of the integrated system of direct currents in the salamander.

———. *Science* 134 (1961): 101. Proof that the direct electrical currents in the salamander are semiconducting in nature.

———. *Journal of Bone and Joint Surgery* 43-A (1961): 643. Original paper on the electrical factors in regenerating limbs of salamanders.

———. *New York State Journal of Medicine* 62 (1962): 1169. First paper proposing the primitive DC control system.

———. *Clinical Orthopedics and Related Research* 73 (1970): 169. The complete electrical control system for the healing of bone fractures.

———. *Journal of Bioelectricity* 1 (1982): 239. This is a review of my own work on electrical controls of regeneration.

———. *Journal of Bioelectricity* 3 (1984): 105. An attempt to "put it all together."

Cohenheim, J. F. *Vorlesungen über allgemeine Pathologie.* Berlin: A. Hirschwald, 1877.

Dusseau, J., *Perspectives in Biology and Medicine* 30 (1987): 345.

Illingworth, C. M. *Journal of Pediatric Surgery* 9 (1974): 853. Report on regeneration of fingertips in children.

Libet, B. *The Sciences* (New York Academy of Science), March–April 1989: 32.

Pedersin-Bjergaard, J., et al. *New England Journal of Medicine* 318 (1988): 1028. The latest medical paper indicating significant cancer-producing risks of treatment with cancer-chemotherapeutic agents.

Rose, S. M., and Wallingford, H. M. *Science* 107 (1948): 457. Disappearance of cancer in regenerating limb of salamander.

Szent-Gyorgyi, A. *Introduction to Submolecular Biology.* New York and London: Academic Press, 1960. The classic in the field. Also see Szent-Gyorgyi's seminal paper "A Study of Energy Levels in Biochemistry." *Nature* 148 (1941): 157.

Thiemann, W., and Jarzak, U. *Origin of Life* 11 (1981): 85.

Wolsky, A. *Growth* 42 (1978): 425. Paper on regeneration and cancer.

Additional references and discussion may be found in Robert O. Becker and Andrew A. Marino, *Electromagnetism and Life* (Albany, N.Y.: State University of New York Press, 1982), and in *Modern Bioelectricity*, edited by A. A. Marino (New York: Marcel Dekker, 1988). The most up-to-date material available.

Author's note: Coley toxin work is still going on. See the series of monographs edited by Helen C. Nauts and published by the Cancer Research Institute in New York, 1975–76. The present work, which centers on specific chemical substances related to Coley's toxin, is reviewed in *Scientific American,* May 1988.

CHAPTER THREE
THE NEW SCIENTIFIC REVOLUTION: THE MAGNETIC CONNECTION

Baker, R. R. "Human Magnetoreception for Navigation." In *Electromagnetic Fields and Neurobehavioral Function*, edited by M. E. O'Conner and R. H. Lovely. New York: Alan R. Liss, 1988.

Blakemore, R. *Science* 190 (1975):377. The first paper describing the magnetic organ in bacteria.

Brown, Frank. Many papers on biological cycles, published in the 1960s and 1970s. Best summary is in *American Scientist* 60 (1972):756.

Cohen, D. *Science* 175 (1972):664. First report of the magnetic encephalogram (magnetic field given off by the brain).

Keeton, W. T. *Proceedings of the National Academy of Science* 68 (1971):102. Keeton's paper proved for the first time that homing pigeons have a magnetic navigation system.

Kolata, G. *Science* 233 (1986):417. This is a scientific news report on the periodicity of human heart attacks at 9 A.M. Original paper was by Muller, J. E., in *New England Journal of Medicine* 313 (1985):1315.

Moore-Ede, M. C., et al. *New England Journal of Medicine* 309 (1983):469 and 309 (1983):530. Two review papers on biological cycles and diseases.

Munz, H. *Journal of Comparative Physiology (A)*, 154 (1984): 33. Sensing of electrical fields by fish.

Phillips, J. B. *Science* 233 (1986):765. Two magnetic receptors in the salamander.

Preslock, J. P. *Endocrine Reviews* 5 (1984):282. Review of functions and mechanisms of pineal gland.

Semm, P. *Nature* 228 (1980):206. First description of magnetic sensitivity of pineal gland.

Sulzman, F. *Science* 225 (1984):232. Report on the space-shuttle study of biological cycles.

Walcott, C. *Science* 205 (1979):1027. First identification of a magnetic organ in the homing pigeon.

———. *Science* 205 (1979):1077.

Walker, M. *Science* 224 (1984):751. Associating magnetic organ with function.

Wever, R. *ELF and VLF Electromagnetic Field Effects*, edited by M. A. Persinger. New York: Plenum Press, 1974.

CHAPTER FOUR
TURNING ON THE BODY'S ELECTRICAL SYSTEM:
MINIMAL-ENERGY TECHNIQUES

Breder, C. D. *Science* 240 (1988):321. Aspects of psychoneuroimmunology.

Fontana, A. *Journal of Immunology.* 129 (1982):2413. Psychoneuroimmunology.

Friedman, H., Becker, R. O., and Bachman, C. H. *Archives of General Psychiatry* 7 (1962):193. DC electrical potentials influenced under hypnosis.

Friedman, H., et al. *Nature* 200 (1963):626. Report on admissions to psychiatric hospitals related to magnetic storms.

Green, E., and Green, A. *Beyond Biofeedback.* San Francisco: Delacorte Press, 1977. The best book on biofeedback, plus much discussion of other phenomena, by *the* authorities.

Hasson, J. *Cancer Research* 18 (1958):267. Report of increased growth of cancer in denervated portions of the body.

Holden, C. *Science* 200 (1978):1316. An excellent review of the mind/cancer relationship; gives an indication of how little has been done since then.

Krippner, S., ed. *Advances in Parapsychological Research.* Vol. 1, *Psychokinesis* (1977); vol. 2, *Extrasensory Perception* (1978); vols. 3 (1982) and 4 (1984), untitled. New York and London: McFarland, Jefferson. All volumes are of great value in that they collect the best available scientific evidence for all aspects of parapsychology.

Norris, P., and Porter, G. *I Choose Life.* Walpole, N.H.: Stillpoint Publishing, 1987. Report of a dramatic outcome to visualization therapy.

Pawlowski, A., and Weddell, G. *Nature*, March 25, 1967: 1234. Increased growth of cancer in denervated portions of the body.

Rauscher, E., and Van Bise, W. Personal communication, 1987.

Riley, V. *Science* 212 (1981):1100. Major paper reporting many mind/hormone relationships.

Sklar, L. S., and Anisman, H. *Science* 205 (1979):513. Report that electrical stimulation of portions of the brain increases the growth of established cancers.

Tromp, S. W. *Psychical Physics.* New York, Amsterdam, London, and Brussels: Elsevier Publishing Co., 1949. A major, and fascinating, book that covers all areas of the relationship between physics and living organisms. Although the book was written some time ago, most of the material is still pertinent.

Visintainer, M. A., et al. *Science* 216 (1982):437. Report on how electrical stimulation of the brain increases the growth of cancers.

CHAPTER FIVE
BUILDING UP THE BODY'S ELECTRICAL SYSTEM:
ENERGY-REINFORCEMENT TECHNIQUES

Davenas, E., et al. (including laboratory director, Benveniste, J.). *Nature* 333 (1988):816. Controversial paper reporting biological effects from exceedingly high dilutions.

Israelachvili, J., and McGuiggan, P. *Science* 241 (1988):795. Report of strong forces at very close distances in liquids.

Maddox, J., Randi, J., and Stewart, W. *Nature* 334 (1988):287. Report by the investigative team on Benveniste's experiments, labeling them "delusions."

Reichmanis, M., et al. *IEEE Transactions on Biomedical Electronics* 22 (1975):533. Our original publication showing a direct relationship between electrical measurements and acupuncture points.

————. *American Journal of Chinese Medicine* 4 (1976):69. Acupuncture points show increased DC electrical conductivity.

————. *IEEE Transactions on Biomedical Electronics* 24 (1977):402. Acupuncture meridians have transmission-line characteristics.

Reilly, D. T., et al. "Is Homeopathy a Placebo Response?" in *Lancet*, October 1986. This is the latest and most convincing report of the clinical effectiveness of homeopathic preparations. It includes a good bibliography as well.

Schoen, A. M., et al. *Seminars in Veterinary Medicine and Surgery (Small Animals)* 1 (1986):224. A good review of veterinary acupuncture and its clinical utility.

Ullman, D. *Homeopathy for the 21st Century.* Berkeley, Calif.: North Atlantic Books, 1988. The latest book on the subject, quite complete and written for the general reader.

Upledger, J. E., and Vredevoogd, J. D. *Craniosacral Therapy.* Seattle: Eastland Press, 1983. Excellent account of manipulation therapy.

Additional information may be obtained from the Homeopathic Education Service, 2124 Kittredge St., Berkeley, CA 94704.

CHAPTER SIX
ADDING TO THE BODY'S ELECTRICAL SYSTEM: HIGH-ENERGY
TRANSFER TECHNIQUES

Akaimine, T. *Report of the Japan Committee of Electrical Enhancement of Bone Healing.* Abstract no. 10, 1981. Report showing that pulsed electromagnetic fields produce an increase in the growth of cancer cells.

Bassett, C. A. L., and Becker, R. O. *Science* 137 (1962):1063. Report that bone has a piezoelectric effect.

Bassett, C. A. L., Pawluk, R., and Becker, R. O. *Nature* 204 (1964):652. Report that DC electrical current stimulates bone growth.

Becker, R. O., Bassett, C. A. L., and Bachman, C. H., in *Bone Biodynamics*, edited by H. Frost. Boston: Little, Brown & Co., 1964. Theory relating piezoelectric effect in bone to its stress-related growth.

Becker, R. O., and Murray, D. G. *Clinical Orthopedics and Related Research* 73 (1970):169. Complete description of the fracture-healing electrical control system.

Becker, R. O., et al. *Clinical Orthopedics and Related Research* 124 (1977):75. Report that very-low-strength negative electrical currents could cause bone growth and healing.

Becker, R. O., and Esper, C. *Clinical Orthopedics and Related Research* 161 (1981):336. Report that low-strength electrical currents stimulate the growth of cancer cells.

Becker, R. O. *Calcified Tissue Research* 26 (1978):93.

————. *Clinical Orthopedics and Related Research* 141 (1979):266. Two papers that summarized the DC control system for human bone growth and the possible side effects of electrical and electromagnetic treatments.

Friedenberg, Z. B., and Brighton, C. T. *Journal of Bone and Joint Surgery* 48A (1566):915. Report that human fractures, like the salamander's regenerating limb, are electrically negative.

Friedenberg, Z. B., et al. *Trauma* 11 (1971):883. First report of the use of negative electrical current to heal a human bone nonunion.

Fukada, E., and Yasuda, I. *Journal of Physiology Society Japan* 12 (1957):1158. First report of piezoelectric effect in bone.

Fukata, H. *Report of the Japan Committee of Electrical Enhancement of Bone Healing.* Abstract no. 8, 1981. Report that DC electrical current increases DNA synthesis in cancer cells.

Guderian, R. H., et al. *Lancet,* July 26, 1986. Report on the high-voltage treatment for snakebite.

Humphrey, C. E., and Seal, E. H. *Science* 130 (1959):388. Report that positive DC electrical currents slow growth of cancer cells.

Iida, H., et al. *Journal of Kyoto Prefecture Medical University* 60 (1956):561. First report that electrical currents could stimulate bone growth.

Long, D. M. *Minnesota Medicine* 57 (1974):195. Confirmed that electrical stimulation of the skin is useful for treating chronic pain.

Martin, Franklin H. *Fifty Years of Medicine and Surgery.* Chicago: Surgical Publishing Co., 1934. Autobiography, containing a description of Dr. Martin's experiments using electricity to treat cancer in the 1880s.

Melzak, R., and Wall, D. W. *Science* 150 (1965):971. Proposed the gate theory of pain.

Noguchi, K. *Journal of Japan Orthopedics* 31 (1957):1. Second report that electrical currents stimulate bone growth.

Nordenström, B. *Biologically Closed Electrical Circuits.* Stockholm: Nordic Medical Publications, 1983. Presents Nordenström's theory and method of electrical treatment of cancer.

Schaubel, M. K., and Habal, M. B. *Archives of Pathology and Laboratory Medicine* 101 (1977):294. Study of the positive DC electrical effect on cancer cells.

Shealy, C. N., et al. *Anesthesia and Analgesia* 45 (1967):489. Electrical stimulation of the tracts in the spinal cord to treat chronic pain.

Shealy, C. N. *Surgical Forum* 23 (1973):419. Electrical stimulation of the skin for treatment of chronic pain.

Sperber, D., et al. *Naturwissenschaften* 71 S (1984):100. Magnetic resonance imaging scanners cause increases in body temperature by acting on the thermoregulatory center in the brain.

Tapio, D., and Hymes, A. C. *New Frontiers in Transcutaneous Electrical Nerve Stimulation.* Minnetonka, Minn.: Lectec Corporation, 1987. Excellent review of TENS use.

CHAPTER SEVEN
THE NATURAL ELECTROMAGNETIC FIELD

Becker, R. O. *New York State Journal of Medicine* 63 (1963):2215. First presentation of the concept that the natural magnetic environment exerts an influence on human behavior.

Cole, F. E., and Graf, E. R. "Pre-Cambrian ELF and Abiogenesis," in *ELF and VLF Electromagnetic Field Effects,* edited by M. A. Persinger. New York: Plenum Press, 1974. Presentation of the theory that the Precambrian magnetic field was related to the origin of life.

Friedman, H., Becker, R. O., and Bachman, C. H. *Nature* 200 (1963):626. Paper reporting statistically significant relationship between magnetic storms and rate of admissions to mental hospitals.

————. *Nature* 205 (1965):1050. Report of significant relationship between cosmic-ray counts and patient behavior in psychiatric hospitals.

Hays, J. *Geological Society of America Bulletin* 82 (1971):2433. Significant relationship between magnetic reversals and species extinctions in radiolaria.

Lanzerotti, L. J. "The Earth's Magnetic Environment." *Sky and Telescope,* October 1988. Popular version of present knowledge of Earth's magnetic-field complexities.

Lerner, E. J. "The Big Bang Never Happened." *Discover,* June 1988. The concept of magnetic rather than gravitational forces as involved in the formation of the universe.

Miller, S. L. *Science* 117 (1953):528. Experiment producing amino acids and peptides from simple compounds; possible chemical origin of life.

Newell, N. D. "Crises in the History of Life." *Scientific American* 208 (1963):77. First report of several major species extinctions in the past.

Sander, Christine. "The Unifying Principle: Magnetic Fields and Life." Master's thesis (anthropology), University of California, Santa Barbara, 1984. An excellent review of data available at the time, particularly magnetic reversals and species extinctions.

Thiemann, W., and Jarzak, U. *Origins of Life* 11 (1981):85. Paper detailing method of producing single form of organic chemical by magnetic field.

Weller, G., et al. *Science* 238 (1987):1361. A recent review of scientific research and present knowledge of the magnetosphere.

CHAPTER EIGHT
MAN-MADE ELECTROMAGNETIC FIELDS

Anderson, B. S., and Henderson, A. K. *Cancer Incidence in Census Tracts with Broadcasting Towers in Honolulu, Hawaii.* Honolulu: State of Hawaii Department of Health, October 27, 1986.

Aurell, E., and Tengroth, B. *Acta Ophthalmologica* 51 (1973):764. Report that microwaves can produce cataract at nonthermal levels and can damage the retina itself.

Becker, R. O. *Medical Electronics and Biological Engineering* 1 (1963):293. Review of all reports on the biological effects of magnetic fields available at that time.

————. *Journal of Bioelectricity* 7 (1988):103. Analysis of the New York State Department of Health Power-Lines Report.

Becker, R. O., and Becker, A. J. *Journal of Bioelectricity* 5 (1986):229. Analysis of the steps not taken by agencies investigating a possible link between the excess of Down's syndrome in Vernon, N.J., and microwaves.

Biological Effects of Nonionizing Electromagnetic Radiation (BENER Digest). Washington, D.C.: Office of the Chief of Naval Research. Published quarterly.

Biological Effects of Power Line Fields. Final report of the New York State Department of Health, Scientific Advisory Panel, July 1, 1987.

Biological Studies of Swine Exposed to 60-Hz Electric Fields (E.A.-4318). Palo Alto, Calif.: Electric Power Research Institute, 1987. Eight-volume report on the Battelle minipig study.

Birge, R., et al. *Journal of the American Chemical Society* 109 (1987):2090. Report that certain chemicals in the retina absorb microwaves to a high degree.

Brown, H. D., and Chattopadhyay, S. K. *Cancer Biochemistry and Biophysics* 9 (1988):295. Review of all available scientific literature on electromagnetic fields and cancer.

Cohen, B. H. *Population Genetics—Studies in Humans,* edited by E. B. Hook and I. H. Porter. New York: Academic Press, 1977. Repeat of the study relating Down's syndrome to radar exposure that failed to find relationship.

Delgado, J. M. R., et al. *Journal of Anatomy* 134 (1982):533. Reports on developmental defects in chick embryos exposed to various ELF frequencies.

Epstein, S. S. *Science* 240 (1988):1043. Reasoned comment on the real incidence of different types of cancer.

Foster, K. R., and Guy, A. W. *Scientific American* 255 (1986):32. This article reports on Guy's experiments but fails to mention crucial aspects that affect the conclusions.

Garfinkel, I., and Savokhan, B. *Annals of the New York Academy of Sciences* 381 (1982):1. Report that the incidence of brain tumors rose between 1940 and 1977.

Goodman, R. *Proceedings of the National Academy of Sciences* 85 (1988):3298.

Heller, J. H., and Teixeira-Pinto, A. A. *Nature* 183 (1959):905. First report on the production of chromosomal abnormalities by fields at 27 MHz.

Hosmer, H. *Science* 68 (1928):327. First report of a bioeffect of radiowaves, with the heating effect observed.

Janes, D. E., et al. *Nonionizing Radiation* 1 (1969):125. Confirmed and extended Heller's report on chromosome effect.

Liboff, A. *Science* 223 (1984):818. Report that a wide range of frequencies in the ELF-VLF range could increase the rate of DNA synthesis in dividing cells.

Lilienfeld, A. M. *Final Report on Contract #6025-619073*. Washington, D.C.: Department of State, 1978. Report on health status of people in American embassy, Moscow, with microwave irradiation by Russians; no problems reported, but data not reported.

Lin, R. S., et al. *Journal of Occupational Medicine* 27 (1985):413. Report on relationship between incidence of brain tumors and occupational electromagnetic exposure.

Manikowska-Czerska, E., Czerska, P., and Leach, W. M. In *Proceedings US-USSR Workshop on Physical Factors—Microwave and Low-Frequency Fields*. Washington, D.C.: National Institute of Environmental Health Sciences, 1985.

Marino, A. A., and Ray, J. *The Electric Wilderness*. San Francisco: San Francisco Press, 1986. Full account of the New York State Public Service Commission hearings on possible health effects of high-voltage power lines.

Marino, A. A., et al. *Experientia* 32 (1976):856. Report from my laboratory on the effects of exposure of three generations of mice to 60-Hz electrical fields.

Microwave News. Newsletter that is the only up-to-date review of all actions in the entire bioeffects area of the whole electromagnetic spectrum. Published ten times per year; available by subscription (Box 1799, Grand Central Station, New York, NY 10163).

Morton, W. E., and Phillips, D. S. *Radioemission Density and Cancer Epidemiology in the Portland Metro Area* (Grant #R-805832, EPA). Triangle Park, N.C.: Health Effects Research Laboratory. Report on relationship between level of FM radiation and leukemia.

Nelson, K., and Holmes, L. *New England Journal of Medicine* 320 (1989):19.

Nordenson, I., et al. *Radiation and Environmental Biophysics* 23 (1984):191. Chromosomal abnormalities in lymphocytes of humans exposed to power-frequency fields.

Nordstrom, S., et al. *Bioelectromagnetics* 4 (1983):91. Genetic defects in offspring of power-frequency workers.

Osborne, S. L., and Frederick, J. N. *Journal of the American Medical Association* 137 (1948):1036. First report of experiments to determine whether microwaves below heating effect could produce cataracts; none were found.

Phillips, J., Winters, W. D., and Rutledge, J. *International Journal of Radiation Biology* 49 (1986):463. Report that exposure to 60-Hz electromagnetic and magnetic fields increases rate of growth of human cancer cells.

Phillips, J. *Immunology Letters* 13 (1986):295. Report that human cancer cells exposed to 60-Hz fields resist destruction by body's killer-type cells.

Phillips, J., et al. *Cancer Research* 46 (1986):239. Report that exposure to 60-Hz fields markedly increases the surface-binding sites of human cancer cells.

Phillips, Richard. Letter to Lawrence W. Herbert, 1 February 1988.

Proceedings of the Ad Hoc Committee for the Review of Biomedical and Ecological Effects of ELF Radiation (SANGUINE Report). Washington, D.C.: Department of the Navy, Bureau of Medicine and Surgery, December 6–7, 1973.

Reichmanis, M., et al. *Physiological Chemistry and Physics* 11 (1979):395. Report done with Dr. F. S. Perry in England on the relationship between suicide and exposure to power-line fields.

Richardson, A. W., et al. *Archives of Physical Medicine* 29 (1948):765. First report that microwaves could produce cataracts without heating.

Salzinger, K. *Biological Effects of Power Line Fields.* Albany, N.Y.: New York State Power-Lines Project Scientific Advisory Panel, 1987.

Sigler, A. T. *Bulletin of Johns Hopkins Hospital* 117 (1965):374. First report of relationship between fathers' military exposure to radar and incidence of Down's syndrome in offspring.

Speers, M. *American Journal of Industrial Medicine* 13 (1988):629. Reports thirteenfold increase in brain tumors among electrical-utility workers.

Spitz, M. R., and Cole, C. C. *American Journal of Epidemiology* 121 (1985):924. Reports significant increase in incidence of brain tumors among children of fathers occupationally exposed to electromagnetic fields.

Steneck, Nicholas. *The Microwave Debate.* Cambridge, Mass.: MIT Press, 1984. Excellent, highly readable account of the microwave issue, from its beginning up to 1984.

Sulzman, F. *Biological Effects of Power Line Fields.* Albany, N.Y.: New York State Power-Lines Project Scientific Advisory Panel, 1987.

Thomas, T. L., et al. *Journal of the National Cancer Institute* 79 (1987):233. Epidemiological study of electromagnetic occupational exposure and brain cancers; significant relationships found.

Wertheimer, N., and Leeper, E. *American Journal of Epidemiology* 109 (1979):273. First report of relationship between exposure to 60-Hz magnetic fields from electric lines and childhood cancer.

Wolpaw, J. *Biological Effects of Power Line Fields.* Albany, N.Y.: New York State Power-Lines Project Scientific Advisory Panel, 1987.

CHAPTER NINE
ELF AND THE MIND/BRAIN PROBLEM

Becker, R. O. Presentation at First International Conference on High-Energy Magnetic Fields, MIT, 1962. First presentation of link between magnetic storms and rates of admission to mental hospitals.

————. *New York State Journal of Medicine* 62 (1962):1169. Review of internal DC electrical system and hypothesis that this could be influenced by external magnetic fields.

————. *New York State Journal of Medicine* 63 (1963):2215. Review of evidence for relationship between geomagnetic field and living organisms.

Campbell, H. J. *Smithsonian*, October 1971. Sensory input normally stimulates the pleasure center of the brain.

Delgado, J. M. R. *Physical Control of the Mind: Toward a Psychocivilized Society*. Vol. 41, *World Perspectives*. New York: Harper & Row, 1969.

Friedman, H., Becker, R. O., and Bachman, C. H. *Nature* 200 (1963):626. Paper reporting significant relationship between incidence of naturally occurring magnetic storms and increases in rates of admissions to mental hospitals.

――――. *Nature* 213 (1967):949. Report that low-strength ELF magnetic fields influence human reaction time.

CHAPTER TEN
LINKING UP INNER AND OUTER FIELDS: MECHANISMS OF ACTION

Achkasova, Y. N. *Journal of Hygiene, Epidemiology, Microbiology, and Immunology* (USSR) 22 (1978):415. Growth of bacteria related to changes in Earth's magnetic field produced by sector-boundary passages on sun.

Bawin, S., and Adey, W. R. In *Proceedings of the National Academy of Sciences* 73 (1976):1999. First report of ELF fields producing calcium efflux from nerve cells.

Becker, R. O. *Journal of Bioelectricity* 3 (1984):105. First proposal that resonance with Earth's natural field controls timing of cell division.

――――. *Journal of Bioelectricity* 4 (1985):133. Theory of relationship of living organisms to ELF fields.

Blackman, C. F., et al. *Bioelectromagnetics* 6 (1985):327. Ca^{++} efflux with ELF field, relationship to Earth's magnetic field.

Cohen, D., et al. *Proceedings of the National Academy of Sciences* 77 (1980):1447. The DC magnetic field from human hair follicles.

Jafary-Asl, et al. *Journal of Biological Physics* 11 (1983):15. First report of nuclear magnetic resonance accelerating growth.

Kimball, C. G. *Journal of Bacteriology* 35 (1938):109. Inhomogeneous DC magnetic fields slow growth of yeast cells.

Liboff, A. In *Interaction between Electromagnetic Fields and Cells*, edited by A. Chibrera, C. Nicolini, and H. P. Schwann. New York: Plenum Press, 1985. Relates Ca^{++} efflux with ELF to Earth's magnetic field.

————. *Journal of Biological Physics* 13 (1985):99. Theory of cyclotron resonance.

Mitchell, J. T., et al. *Physiological Chemistry and Physics* 10 (1978):79. Report from my lab on the production of abnormal chromosomes in cancer cells exposed to DC electrical fields.

Poenie, M., et al. *Science* 233 (1986):886. Report that calcium ions increase abruptly within cells at onset of anaphase in cell division.

Thomas, J. R., Schrot, J., and Liboff, A. *Bioelectromagnetics* 7 (1986):349. Production of passive behavior in mice by means of lithium cyclotron resonance.

Walker, J. *Scientific American* 256 (February 1987):134.

CHAPTER ELEVEN
THE NEW PLAGUES

Albert, E., and Sherif, M. *Electromagnetic Fields and Neurobehavioral Function,* edited by M. O'Connor and R. Lovely. New York: Alan R. Liss, 1988.

Barnes, D. E. M. *Science* 243 (1989):171. Review of recent findings on Fragile *X* chromosome syndrome.

Bergstrom, W., and Hakanson, D. Personal communication.

Calne, D. *Canadian Journal of Neurological Science* 14 (1987):303. Report of onset of Parkinson's disease occurring at earlier ages and associated with clustering and familial incidence.

Cicerone, R. J. *Science* 237 (1987):35. Review of current data on atmospheric ozone.

Courchesne, E., et al. *New England Journal of Medicine* 318 (1988):1349. Report that autistic children have anatomical defect in cerebellum. See also an editorial on this condition in the same issue.

Eisenberg, L. *New England Journal of Medicine* 315 (1986):705. Editorial concerning increase in incidence of adolescent suicide.

Fitzpatrick, T. B. Interview with Janis Johnson, as quoted in *USA Today,* 4 March 1985. Epidemic of melanoma in developed countries.

Green, M. H., et al. *New England Journal of Medicine* 312 (1985):91. Complete review article on malignant melanoma.

Hannson, H. A.; Albert, E. N. Separate chapters in *Electromagnetic Fields and Neurobehavioral Function*, edited by M. E. O'Conner and R. H. Lovely. New York: Alan R. Liss, 1988. Both report the production of anatomical defects in the cerebellums of animals exposed immediately after birth to abnormal electromagnetic fields.

Hembree, D., and Henry, S. *San Diego Union*, 5 January 1987. News report of epidemic of hypersensitivity syndrome in Silicon Valley.

Holden, C. *Science* 233 (1986):839. Review of status of adolescent suicide.

Holmes, G. P., et al. *Annals of Internal Medicine* 108 (1988):387. Report on the Centers for Disease Control analysis of chronic-fatigue syndrome.

Katzman, R. *New England Journal of Medicine* 314 (1986):964. Complete, up-to-date review of Alzheimer's disease.

Lawrence Livermore National Laboratory. *Magnetic-Field Exposure Guidelines, 1986.* Livermore, Calif.: LLNL, 1986.

Lewin, Roger. *Science* 237 (1987):978. Interview with Dr. Donald Calne regarding increasing incidence of Parkinson's disease.

Lyle, D. B., et al. *Bioelectromagnetics* 9 (1988):303. Report that T-cell lymphocycte cytotoxic function is depressed by exposure to 60-Hz fields.

Marx, J. *Science* 217 (1982):618. Report of "mystery disease" called AIDS.

O'Leary, C. *Sunday Mirror*, 29 January 1989.

Overbaugh, J., et al. *Science* 239 (1988):906. Report that through molecular cloning techniques, the feline leukemia virus was changed into a virus that produced immunodeficiency syndrome, rather than leukemia, in cats.

Regier, D., as quoted by Jeffrey Fox. *Science* 226 (1984):324. Report of National Institute of Mental Health survey indicating increase in mental disease among general population associated with marked increase among persons below age of forty-five.

Reynolds, P., and Austin, D. *Western Journal of Medicine* 142 (1985):214. Report of epidemic of melanoma at Lawrence Livermore National Laboratory.

St. George-Hyslop, P. *Science* 238 (1987):664. Review of chromosomal abnormalities associated with Alzheimer's disease.

Strauss, S. E., et al. *New England Journal of Medicine* 319 (1988):1692. National Institutes of Health acyclovir trials on chronic-fatigue syndrome, with report that patient's mood factors change clinical status.

Sturner, W. *New Scientist* 10 (1987):37.

Teravanian, H. *Canadian Journal of Neurological Science* 13 (1986):317. Report of onset of Parkinson's disease occurring at earlier ages and associated with clustering and familial incidence.

Weissman, M. M. *Science* 235 (1987):522. Comments on increased incidence of mental disease in U.S. population.

CHAPTER TWELVE
THE RISK AND THE BENEFIT: WHAT YOU CAN DO

DeMatteo, R. *Terminal Shock*. Toronto, Canada: NC Press, 1985.

DeSilva, Bruce. Interview with Joe Bearskin. *Watertown Times*. 4 January 1989.

Goldhaber, M. K., et al. *American Journal of Industrial Medicine* 13 (1988):695.

Lovins, A. R. *Science* 229 (1986):914.

Marlay, R. C. *Science* 226 (1984):1277.

Microwave News (P.O. Box 1799, Grand Central Station, New York, NY 10163). Vol. 6, March-April 1986; vol. 7, July-August 1987; vol. 8, March-April 1988 and May-June 1988. Reports of most recent work on health problems associated with computers and VDTs.

Mikolajczyk H., et al. In *International Scientific Conference on Work with Display Units*, edited by B. Knave and P. G. Wideback. Amsterdam: Elsevier Scientific Publications, 1987. Report of study in which rats were exposed to commercial TV sets.

Milham, S. *Environmental Health Perspectives* 62 (1985):297. Epidemiological study of persons occupationally exposed to all types of man-made electromagnetic radiation.

———. *Lancet*, 6 April 1985. Study of mortality rates among amateur radio operators.

Union of Concerned Scientists (26 Church Street, Cambridge, MA 02238). Voter-information paper on nuclear and energy policy, 1987.

Wertheimer, N., and Leeper, E. *Bioelectromagnetics* 7 (1986):13. Electric blanket/miscarriage study.

Equipment for measuring electromagnetic fields can be obtained from the following companies:

Integrity Electronics and Research
558 Breckenridge Street
Buffalo, NY 14222
    Excellent, inexpensive 60-Hz magnetic-field monitor, usable by the average homeowner.

Monitor Industries
6112 Four Mile Canyon
Boulder, CO 80302
    Excellent, professional-grade 60-Hz magnetic-field monitor.

Narda Microwave Products
435 Moreland Road
Hauppauge, NY 11788
    Maker of complete range of sophisticated monitoring equipment for rental or purchase. For use by knowledgeable persons only.

Holaday Industries
14825 Martin Drive
Eden Prairie, MN 55344
    Same characteristics as Narda Microwave Products devices.

APPENDIX
THE HIDDEN HAND ON THE SWITCH: MILITARY USES OF THE ELECTROMAGNETIC SPECTRUM

Steneck, Nicholas H. *The Microwave Debate*. Cambridge, Mass.: MIT Press, 1984. An excellent historical review of the development of microwave technology and its relationship to the military. Also presents some details of various secret projects as they relate to the question of hazard.

Tyler, Paul E. "The Electromagnetic Spectrum in Low-Intensity Conflict." In *Low-Intensity Conflict and Modern Technology*, edited by Lt. Col. David J. Dean, U.S.A.F. Center for Aerospace Doctrine, Research, and Education, Maxwell Air Force Base, Ala.: Air University Press, 1986. Low-intensity conflicts are "small" wars that do not involve the use of nuclear weapons. This article discusses the direct use of electromagnetic fields against personnel.

"Walter Reed's Microwave Research Department: Its History and Mission [Part 1 of two parts]." In *Bioelectromagnetics Society Newsletter*, January-February 1989. This unattributed article is ostensibly about safety standards, but it is primarily a discussion of high-power pulsed microwave as it relates to antipersonnel use. The Bioelectromagnetics Society has strong links to the military establishment and is considered to be an authoritative source.

INDEX